Editorial Policy

for the publication of monographs

In what follows all references to monographs are applicable also to multiauthorship volumes such as seminar notes.

§ 1. Lecture Notes aim to report new developments - quickly, informally, and at a high level. Monograph manuscripts should be reasonably self-contained and rounded off. Thus they may, and often will, present not only results of the author but also related work by other people. Furthermore, the manuscripts should provide sufficient motivation, examples, and applications. This clearly distinguishes Lecture Notes manuscripts from journal articles which normally are very concise. Articles intended for a journal but too long to be accepted by most journals usually do not have this "lecture notes" character. For similar reasons it is unusual for Ph.D. theses to be accepted for the Lecture Notes series.

§ 2. Manuscripts or plans for Lecture Notes volumes should be submitted (preferably in duplicate) either to one of the series editors or to Springer-Verlag, New York. These proposals are then refereed. A final decision concerning publication can only be made on the basis of the complete manuscript, but a preliminary decision can often be based on partial information: a fairly detailed outline describing the planned contents of each chapter, and an indication of the estimated length, a bibliography, and one or two sample chapters - or a first draft of the manuscript. The editors will try to make the preliminary decision as definite as they can on the basis of the available information.

§ 3. Final manuscripts should be in English. They should contain at least 100 pages of scientific text and should include

- a table of contents;
- an informative introduction, perhaps with some historical remarks: it should be accessible to a reader not particularly familiar with the topic treated;
- a subject index: as a rule this is genuinely helpful for the reader.

Lecture Notes in Statistics 98

Edited by P. Diggle, S. Fienberg, K. Krickeberg, I. Olkin, N. Wermuth

Philippe Barbe and Patrice Bertail

The Weighted Bootstrap

Springer-Verlag
New York Berlin Heidelberg London Paris
Tokyo Hong Kong Barcelona Budapest

Philippe Barbe
CNRS, Laboratoire de Statistiques
et Probabilité,
Université Paul Sabatier
118, route de Narbonne
31 062 Toulouse Cedex
France

Patrice Bertail
INRA-CORELA,
63, Bd de Brandebourg
94 205 Ivry sur Seine Cedex
France

Library of Congress Cataloging-in-Publication Data Available
Printed on acid-free paper.

Camera ready copy provided by the editor.

9 8 7 6 5 4 3 2 1

ISBN-13: 978-0-387-94478-4 e-ISBN-13: 978-1-4612-2532-4
DOI: 10.1007/978-1-4612-2532-4

TABLE

Acknowledgements

Introduction 1

I.1) Introduction 1

I.2) Some connected works 3

I) Asymptotic theory for the generalized bootstrap of statistical differentiable functionals 9

I.1) Introduction 9

I.2) Fréchet-differentiability and metric indexed by a class of functions 11

I.2.1) Differentiability assumptions 11

I.2.2) The choice of the metric 14

I.2.3) Rate of convergence of the weighted empirical process indexed by a class of functions 17

I.3) Consistency of the generalized bootstrapped distribution, variance estimation and Edgeworth expansion 19

I.3.1) Consistency of the generalized bootstrapped distribution 19

I.3.2) The generalized bootstrap variance estimator 20

I.3.3) Edgeworth expansion of the studentized functional 23

I.3.4) Inverting Edgeworth expansion to construct confidence intervals 25

I.4) Applications 29

I.4.1) The mean 29

I.4.2) M-estimators 30

I.4.3) The probability of being censored 30

I.4.4) Multivariate V-statistics 31

I.5) Some simulation results 32

II) How to choose the weights 45

II.1) Introduction 45

II.2) Weights generated from an i.i.d. sequence : almost sure results 46

II.3) Best weights for the bootstrap of the mean via Edgeworth expansion 48

II.3.1) Second order correction 48

II.3.2) Coverage probability 52

II.4) Choice of the weights for general functional via Edgeworth expansion 53

II.4.1) Edgeworth expansion up to $o(n^{-1})$ for a third order differentiable functional 54

II.4.2) Edgeworth Expansion up to $o(n^{-1})$ for the weighted version 55

II.5) Coverage probability for the weighted bootstrap of general functional 59

II.5.1) Derivation of the coverage probability 59

II.5.2) Choosing the weights via minimization of the coverage probability 66

II.5.3) Simulation results 70

II.6) Conditional large deviations 72

II.7) Conclusion 76

III) Some special forms of the weighted bootstrap 77

III.1) Introduction 77

III.2) Bootstrapping an empirical d.f. when parameters are estimated or under some local alternatives 77

III.3) Bootstrap of the extremes and bootstrap of the mean in the infinite variance case 85

III.4) Conclusion 90

IV) Proofs of results of Chapter I 93

IV.1) Proof of Proposition I.2.1 93

IV.2) Proof of Proposition I.2.2 95

IV.3) Proof of Theorem I.3.1 98

IV.4) Some notations and auxilliary lemmas 99

IV.5) Proof of Theorem I.3.2 101

IV.6) More lemmas to prove Theorem I.3.2 103

IV.7) Proof of Theorem I.3.3 107
IV.8) Proof of Theorem I.3.4 114
IV.9) Proof of Theorem I.3.5 118

V) Proofs of results of Chapter II 119
V.1) Proofs of results of section II.2 119
V.2) Proof of Formula (II.3.2) 125
V.3) Proof of Proposition II.4.1 126
V.4) Proof of (II.5.6) 134
V.5) Proof of (II.5.9) 137
V.6) Proof of (II.5.10) 138
V.7) Proof of (II.5.11) 143
V.8) Proof of Theorem II.6.2 143

VI) Proofs of results of Chapter III 145
VI.1) Proof of Theorem III.1.1 145
VI.2) Proof of Theorem III.1.2 147
VI.3) Proof of Theorem III.2.1 148
VI.4) Proof of Theorem III.2.2 151

Appendix 1 : Exchangeable variables of sum 1 153
Appendix 2 : How to express $E_W Q_{W,n}^p$ as a sum of U-statistics 159
Appendix 3 : Derivation of $E_W Q_{W,n}^2$ and S_n^2 163
Appendix 4 : Calculation of $E_W Q_{W,n}^3$ 169
Appendix 5 : Finite sample asymptotic for the mean and the bootstrap mean estimator 175
Appendix 6 : Weights giving an almost surely consistent bootstrapped mean 177
Appendix 7 : Differentials of the functional $T^{(1)}(X,P)$ 181
Appendix 8 : Edgeworth expansions up to $o(n^{-1})$ for a third order differentiable statistical functional and its studentized form 183

References 199
Notation index 215
Author index 223
Subject index 229

Acknowledgements

It is a great pleasure for the authors to thank many persons who contributed to this monograph by various ways. This work begun in 1991. By the end of the fall 1991, a very preliminary draft of the first chapter was written. While one of us was visiting David Mason for the first Franco-American semester organized by the Center for New Wave Statistics (CNWS, Newark, Delaware), we showed to David this preliminary draft. Thanks David for your great hospitality and your advice to extend a 'too long' paper into a monograph. We followed this advice and started to improve the draft. Then, one of us visited Paul Janssen at the Limburgs Universitair Centrum, where we learned more about the generalized bootstrap, and continued in a very friendly department to improve the first chapter. Afterwards, Erich Haeusler hosted one of us at the University of Giessen, even providing his own desk to his guest enabling him to finish the final version of the chapter 1 in a short time. David Mason came there too for a while. All of us had some interesting discussions on the bootstrap and the question of the choice of the weights. During this visit, Erich worked out the result on coverage probability used here in chapter 2 as a starting point, and stimulated us to get the Edgeworth expansion for a general functional. Finally came the second Franco-American semester (fall 1992) where David Mason and Vince LaRiccia hosted us beautifully at the CNWS where they organized two great seminar series. We wrote the remainder of our monograph there. The Edgeworth expansion for general functionals was worked out in F. Barron's house (the very pleasant official French house of the Franco-American semester).

We are very grateful to Lancelot James and Armelle Guillou for their careful reading of the original manuscript. We also thank M. Gilchrist and the referees for their comments and suggestions which greatly improved the presentation of the monograph.

Finally, we would also like to thank the CREST and the INRA for their material and financial supports.

INTRODUCTION

1) Introduction

In 1979, Efron introduced the bootstrap method as a kind of universal tool to obtain approximation of the distribution of statistics. The now well known underlying idea is the following : consider a sample $X_1, \ldots, X_n$ of independent and identically distributed (i.i.d.) random variables (r.v.'s) with unknown probability measure (p.m.) P. Assume we are interested in approximating the distribution of a statistical functional $T(P_n)$, the empirical counterpart of the functional $T(P)$, where $P_n := n^{-1} \Sigma_{i=1}^{n} \delta_{X_i}$ is the empirical p.m. Since in some sense P_n is close to P when n is large, if one samples $X_1^*, \ldots, X_{m_n}^*$ i.i.d. from P_n and builds the empirical p.m. $P_{m_n}^* := m_n^{-1} \Sigma_{i=1}^{m_n} \delta_{X_i^*}$, then the behaviour of $T(P_{n,m_n}^*)$ conditionally on P_n should imitate that of $T(P_n)$ when n and m_n get large.

This idea has lead to considerable investigations to see when it is correct, and when it is not. When it is not, one looks if there is any way to adapt it.

More than 10 years later, Mason and Newton (1992) proved the consistency of a general version of the bootstrap, which somehow goes back to Efron's (1982, Chapter 6) resampling plans. As was noticed by Efron (1979), $P_{n,m_n}^* \stackrel{d}{=} \Sigma_{i=1}^{n} W_{i,n} \delta_{X_i}$ where $(m_n W_{1,n}, \ldots, m_n W_{n,n})$ is a random vector with $\mathcal{M}$ultinomial(m, 1/n, ..., 1/n) distribution. Mason and Newton (1992) proved that under fairly general conditions, if one only assumes that the $W_{i,n}$'s are exchangeable then the main properties of Efron's bootstrap remain for some smooth functionals.

Hence, all kinds of randomly weigthed empirical p.m. lead to the same consistency results. So, there is a priori no reason to prefer Efron's scheme to almost any arbitrary random weighted scheme. However, it appears as Haeusler (1992) suggests, that one can do better than Efron's bootstrap if one chooses properly the weights $W_{i,n}$, as a function of the sample.

This approach leads to two basic questions :

1) How well does the generalized bootstrap work ?

2) What are the differences between all the different weighted schemes ?

The aim of this research monograph is to start to answer these two questions. We tried to make the proofs as detailed as possible. However, some lengthy (but rather easy, if well undertaken!) calculations are put in separate appendices to avoid too many breaks during the proofs.

In the first chapter, we investigate the weighted bootstrap of statistical functionals and look for some general regularity conditions under which the generalized bootstrap may be used.

In the second chapter we try to give some information concerning the practical choice of the weights and the differences between all these random weighted methods in the regular cases investigated in the first chapter. It turns out that there is no really nice answer. We cannot say exactly for a given problem how the $W_{i,n}$'s must be chosen. But we can give reasonable indications which are in fact extremely precise if we restrict the distribution function (d.f.) of the $W_{i,n}$'s to be among the standard d.f. that statisticians commonly use. However, the choice of the weights depends essentially on what one considers to be important. Accuracy of the estimation of the entire distribution of the statistic, accuracy of a confidence interval, accuracy in a large deviation sense, accuracy for finite sample size? Some of these criterions may not be satisfied at the same time.

The behaviour of these randomly weighted methods is quite complicated when analyzed with respect to various criterions. For instance, if we consider the coverage probability confidence intervals, as in Haeusler (1992), the possibility to improve substantially Efron's bootstrap may depend on some characteristics of the underlying distribution of the data, but also upon the level of the confidence interval we want to construct!

In the third chapter, we look at some nonregular cases which require a drastic modification of the bootstrap. It is very far from being exhaustive: our aim is more to give some indications on the choice of the weights in some cases of pratical interest, than to establish a definitive theory. The general idea is that one must be very careful in using the bootstrap if the usual assumptions are not satisfied.

Theorems, propositions, formulas, etc. are labelled N.p.q where N is the number of the chapter, p the number of the section in chapter N , and q is the number of the theorem, proposition, formula in section N.p . Within chapter N , we refer to theorem, proposition, formula by p.q .

2) Some connected works

Let us now survey briefly some previous works and motivate our work. In very nice problems, the statistic $T(P_n)$ is well behaved and one can calculate explicitly its distribution. For instance if P is the standard Gaussian probability measure (p.m.) on $\mathbb{R}$ and $T(P) = \int x dP$ is the mean or $T(P) = \int x^2 dP - \left(\int x dP\right)^2$ is the variance, the exact distribution of $T(P_n)$ is known. If P is not specified to be normal, the distribution of the empirical mean or variance can seldom be expressed on a closed form. However, for many statistics $T(P_n)$ a central limit theorem is known, and therefore one knows an approximation of the distribution of $T(P_n)$ when n gets large. Typically, replacing the exact distribution of $T(P_n)$ by its weak limiting behaviour causes an error of order $n^{-1/2}$. The asymptotic approximation can be refined through an Edgeworth expansion (see Pfanzagl (1985), Bhattacharya and Rao (1986), Bhattacharya and Denker (1990)) or a saddlepoint type approximations (see e.g. Barndorff-Nielsen (1988), Reid (1988), Barndorff-Nielsen and Cox (1989), Field and Ronchetti (1990), Ghosh (1994) and references therein). Unfortunately, although they have been investigated over years, high order asymptotics are not broadly applicable, at least in a nonparametric setting. They often require very strong uniformity assumptions on the functional and the underlying distribution.

The resampling methods, among those the Jackknife (Quenouille (1949)) and the bootstrap (Efron (1979)) were originally introduced to estimate the bias and the dispersion of an estimator in nonparametric situations (see e.g. Efron and Gong (1983), Efron and Tibshirani (1986, 1993)). The general principle is to repeatedly use the original sample to obtain several values of the statistics and calculate their mean and variance. It has been emphasized that resampling methods have been made valuable in practice by advances in computer technology (Diaconis and Efron (1983)).

Recall here that what is usually called Efron's bootstrap consists in drawing $X_1^*, \ldots, X_n^*$ i.i.d. from an estimator $\hat{P}_n$ of the p.m. P, that is to choose a resampling size equal to the original sample size. When the data are real valued, $\hat{P}_n$ is equivalent to a cumulative distribution function $\hat{F}_n$, estimating the true distribution function F. Many choices are possible for $\hat{F}_n$. One may choose the empirical distribution function F_n, leading to the well known naive bootstrap.

The empirical distribution function has the drawback to be discrete. In some cases it may be more interesting to consider a smooth estimator of F.

It is then called the smooth bootstrap (Efron (1982), Silverman and Young (1987), Young (1988), DiCiccio, Hall and Romano (1989)). Many authors (e.g. Falk and Reiss (1989), Hall and Martin (1987, 1989)) stressed the advantages of using a smooth bootstrap in problems related to quantile and density estimations.

Finally, if the original model is parametric $\{ F_\theta : \theta\in\Theta \}$, and $\hat\theta_n$ is an estimator of the true parameter, then one may choose $\hat F_n = F_{\hat\theta_n}$, leading to the parametric bootstrap (Efron (1982, 1985, 1987)).

It is not always possible to calculate explicitly the distribution of the statistic under the p.m. $\hat P_n$. That is why one uses simulation techniques. Hence, we should distinguish two levels :

- a methodological level which consists in replacing P by $\hat P_n$.
- a practical level which consists in applying Monte Carlo techniques. This justifies the expression computer intensive methods associated with the bootstrap (Diaconis and Efron (1983)).

The generalized bootstrap uses a different heuristic. It consists mainly in generating a random p.m. $P_{W,n}$ from the empirical one and to approximate the distribution of $T(P_n)$ by that of $T(P_{W,n})$ conditioned on P_n. The p.m. $P_{W,n}$ is of the form $\Sigma_{1\le i\le n} W_{i,n}\delta_{X_i}$ for some random weights $W_{i,n}$. As noted at the beginning, this generalizes the classical bootstrap where the weights $nW_{i,n}$ are multinomial and count the number of time that X_i is resampled. It also includes the Bayesian bootstrap of Rubin (1981), Boos and Monahan (1986) and Lo (1987,1991) as well as other weighted methods proposed by Tu and Zheng (1988), Weng (1989), James (1993) or the wild bootstrap of Härdle (1989), Härdle and Marron (1991), Mammen (1992, 1993).

Let us now review some known results on the bootstrap of the empirical mean which has been studied in great details in the literature (among others, Bickel and Freedman (1981), Singh (1981), Efron (1982), Athreya (1987), Knight (1989), Hall (1986, 1988 a,b, 1990)). Though bootstrapping the mean is not very useful, the related results are very important since they strongly support the method. Moreover, they can be extended to asymptotically linear statistics, which may be of some interest.

In the real valued case, assuming that $E X_i^2 < \infty$, Bickel and Freedman (1981) showed that $n^{1/2} (\overline{X}_n^* - \overline{X}_n)$ converges to the limiting distribution of the empirical mean $n^{1/2} (\overline{X}_n - EX_1)$, where $\overline{X}_n$ is the empirical mean and $\overline{X}_n^*$ is the bootstrapped mean. The convergence here is the weak convergence conditionally on the data, almost surely along the sequence $(X_i)_{i\ge1}$. It means that

$$\lim_{n\to\infty} P\left(n^{1/2} (\overline{X}_n^* - \overline{X}_n) \le x \mid X_1,\dots,X_n \right)$$

$$= \lim_{n\to\infty} P\left(n^{1/2} (\overline{X}_n - EX_1) \le x \right) \quad \text{a.s.} ,$$

and shows that the bootstrap gives an approximation of the distribution of the mean. Künsch (1984) generalizes these results to general stationary processes, by resampling blocks of observations.

Giné and Zinn (1989) proved that $E X_1^2 < \infty$ is a necessary and sufficient condition for the bootstrap of the mean (with resampling size n) to work almost surely. When the central limit theorem holds with infinite variance, consistency in probability along the sequence $(X_i)_{i\ge1}$ may hold (Knight (1989), Arcones and Giné (1989)). Related results may be found in Hall (1990), Deheuvels, Mason and Shorack (1993).

The consistency of the generalized bootstrap for the mean has been proved in Mason and Newton (1992). Their approach is, up to our knowledge, the most general one and makes clear that all what is needed for a resampling method to work is some reasonable exchangeable weights. See appendix 1 for some properties of such weights.

Bickel and Freedman (1980) and Singh (1981) investigated Edgeworth expansions and provided a set of results showing that the bootstrapped distribution of the studentized mean performs better than the normal approximation assuming, among other things, that $E |X_1|^3 < \infty$. Indeed, they showed that the bootstrap distribution may be interpreted as an empirical Edgeworth expansion which takes into account the asymmetry of the underlying distribution (see also Bhattacharya and Qumsiyeh (1989)). Under a Cramér condition and $E X_1^6 < \infty$, Singh (1981) showed that

$$n (\log\log n)^{-1/2} \| P(n^{1/2}(\overline{X}_n - E X_1)/\sigma \le x) -$$

$$P(n^{1/2}(\overline{X}_n^* - \overline{X}_n)/\hat{\sigma}_n \le x \mid X_1,\dots,X_n) \| < \infty \quad \text{a.s.}$$

where $\hat{\sigma}_n^2$ is the natural estimator of the variance σ^2 and $\| \ \|$ the sup norm.

Bhattacharya and Qumsiyeh (1989) proved that the bootstrap approximation is even more accurate than Edgeworth expansions for L_p norms between distribution functions. These results have been generalized by Babu and Singh (1983) and Bhattacharya (1987) to statistics that are smooth functionals of the empirical moments (see Hall (1992) for further references).

For some special types of weights, a generalized bootstrap version of the results of Singh (1981) may be found in Haeusler, Mason and Newton (1993). One aim of this monograph is to investigate further how good is the generalized bootstrap approximation according to the various possible choices of the weights (see chapter 2).

We now survey some results for the bootstrap of various functional and mention their generalized bootstrap versions when they exist.

Bickel and Freedman (1981) proved that the bootstrap of the empirical distribution function is consistent. Using obvious notations, it means that

$$\lim_{n\to\infty} P\big(n^{1/2}(F_n^*(.) - F_n(.)) \in A \mid X_1,..X_n\big) = P\big(B\circ F(.) \in A \big)$$

a.s. where A is any continuity set for the p.m. of the Brownian bridge $B(.)$ (see also Shorack (1982)). The generalized bootstrap analogue is in Mason and Newton (1992).

Similar result exists for the quantile process (Bickel and Freedman (1981), Singh (1981), Babu and Singh (1983), Falk (1988), Falk and Reiss (1989)). It should be noticed that the accuracy of the bootstrap approximation for a fixed smoothed quantile is only of order $n^{-1/4}$ (Falk and Reiss (1989)). Related results are given by Ghosh, Parr, Singh, Babu (1984), Liu and Singh (1987), Dohmann (1989), Hall and Martin (1989, 1991), Falk and Kaufman (1991).

In the real valued case, Csörgő and Mason (1989) gave a bootstrap analogue of the weighted approximation of Csörgő, Csörgő, Horváth and Mason (1986), Mason and Van Zwet (1987). This yields to a metatheorem assessing that any result proved using only linear bounds and a weighted approximation of the empirical process has a consistent bootstrap analogue. Here also, a generalized bootstrap version exists and follows from the weighted approximation for exchangeable processes of Einmahl and Mason (1992) (see Appendix 1).

The bootstrap of the abstract empirical process à la Dudley (1984) or Pollard (1984) has been worked out by Beran and Millar (1986), Beran, LeCam and Millar (1987), Gaenssler (1987), Gaenssler and Stute (1987), Giné and Zinn (1990, 1991), Sheehy and Wellner (1992). These results show the wide applicability of the bootstrap. The generalized bootstrap version has been found by Praestgaard (1990, 1992) and Praestgaard and Wellner (1993). This approach will be proved to be usefull in studying the weighted bootstrap of very general functionals (see chapter I).

Higher order properties (than just consistency) of the bootstrap of general functional requires a bootstrap theory of Von Mises (1947) statistics or similarly U-statistics. Bickel and Freedman (1981) proved the consistency of the bootstrap of quadratic non-degenerate U-statistics. A very general statement is in Bretagnolle (1983) who showed that in the degenerate case, one needs to undersample (that is typically to choose bootstrap samples of size $o(n/(\log n)^{1+\varepsilon})$, for some $\varepsilon>0$, as $n \longrightarrow \infty$). Related works are in Dehling and Mikosh (1992), Arcones and Giné (1993), Dehling, Denker and Woyczynski (1990) and Helmers, Janssen and Veraverbeke (1992). Hušková and Janssen (1993a, b), Janssen (1993) and Aerts and Janssen (1994) investigated similar results for the generalized bootstrap in the nondegenerate case as well as in the degenerate one.

Further results and higher order properties for the usual bootstrap of U-statistics can be found in Helmers (1991), Athreya, Ghosh, Low and Sen (1984). Among other things, higher order properties for the generalized bootstrap of nondegenerate U-statistics may be deduced from our results.

Beran (1984), Lohse (1984, 1987), Parr (1985), Beran and Millar (1986), Gill (1989), Pons and Turkheim (1989) obtained the consistency of the bootstrap for some differentiable statistics. These results may be applied to some L- and M-estimators (see Sen (1988) for a survey and references therein).

At this stage, we should mention the recent works on undersampling ot bootstrapping without replacement by Politis and Romano (1992), Bertail (1994) and Bickel (1994) that allow to obtain the asymptotic first or second order correctness in a very general setting. It should be noticed that this undersampling scheme is a special case of the generalized bootstrap.

The main aim of this monograph is to investigate the generalized bootstrap and its high order properties for differentiable functionals. Our regularity assumptions are rather weak, and even for the regular bootstrap turn out to be even weaker than those previously considered.

This short review is very far from being exhaustive and surveys only results that are closely related to our investigation of the weighted bootstrap. Among the topics omitted are nonparametric density estimation, linear models, time series, survey sampling, spatial statistics, extreme values, survival analysis ... Among the huge litterature, we refer to the recent books of Hall (1992), Efron and Tibshirani (1993), Mammen (1993) and the volume edited by LePage and Billiard (1992) for further references and different aspects of the bootstrap methodology.

In the same spirit, let us say few words on bootstrapped confidence intervals. Many papers are devoted to this aspect of the bootstrap. The reader may refer to Abramovitch and Singh (1985), Efron and Tibshirani (1986), Hall (1986, 1988b, 1992), Beran (1987, 1988), Efron (1987), DiCiccio and Tibshirani (1987), DiCiccio and Romano (1988), Bickel (1992). Beran (1987) and Hall (1986, 1988b, 1992) show that better intervals are obtained by bootstrapping a near-pivotal or asymptotically pivotal statistics. Let $S(P)$ be the asymptotic variance of $T(P_n)$. Then, $n^{1/2}(T(P_n) - T(P))/S(P)$ is often asymptotically pivotal. However, $S(P)$ is generally unknown and estimators are often difficult to obtain when $T(.)$ is nonlinear. In some specific cases, generally the mean or regular functions of moments, Singh (1981), Hall (1986, 1988b), Beran (1987, 1988), Bhattacharya (1987) have shown that the bootstrap allows one to construct second order correct confidence intervals (see the surveys of DiCiccio and Romano (1988) and Hinkley (1988)). One of the aims of this monograph is to obtain similar results for the weighted bootstrap.

Nevertheless, it may be difficult to obtain a suitable estimator of the asymptotic variance. Many estimators of the variance have been proposed in the literature. Among them, the bootstrap variance estimator (Efron (1979, 1981, 1982), Efron and Gong (1983), Efron and Tibshirani (1986)), the jackknife variance estimator (Shao (1989, 1991) and Shao and Wu (1989))... In this monograph, among other things, we investigate the convergence of a generalized bootstrap variance estimator. We show that this estimator is generally efficient. Some specific results on the bootstrap variance estimator of quantiles have been obtained by Ghosh, Parr and Singh (1983), Hall and Martin (1987, 1991) who show that the behaviour of the bootstrap of the quantile is quite different from that of the mean.

A way to construct a second order correct confidence interval is to obtain the explicit Edgeworth expansion of $n^{1/2}(T(P_n) - T(P))/S_n$ (as done by Hall and Martin (1991) for quantiles) and invert it thanks to "good" estimators of the coefficients involved in the expansion (see Hall (1983), Abramovitch and Singh (1985)). Another way to achieve second order order correctness is to bootstrap the studentized statistics. In this monograph, we propose a third way, between the two above, which consists in choosing suitably the weights in the generalized bootstrap. The main result is that this allows one to reach third order correctness in a rather general setting for regular statistics.

CHAPTER I

ASYMPTOTIC THEORY FOR THE GENERALIZED BOOTSTRAP OF STATISTICAL DIFFERENTIABLE FUNCTIONALS

1) Introduction

Let T be a statistical functional defined on a space $\mathcal{P}$ of probability measures (p.m.'s) on a locally compact Banach space B . Let $X, X_1, \ldots, X_n$ be a sequence of independent and identically distributed (i.i.d.) random variables (r.v.) with common probability $P\in\mathcal{P}$, and let us define

$$P_n := n^{-1} \Sigma_{i=1}^n \delta_{X_i} ,$$

the empirical measure, where δ_{X_i} denotes the Dirac measure at X_i . When T is smooth in a neighborhood of P , a natural estimator of $T(P)$ is its empirical counterpart $T(P_n)$ (see Von Mises (1947), Huber (1981), Manski (1988)).

To define the generalized bootstrapped empirical measure of Mason and Newton (1993), consider $W_n := \{ W_{i,n} : 1\leq i\leq n, n\geq 1 \}$ a triangular array of row exchangeable r.v. with joint distribution $\mathcal{W}_n$ on the n-th row, and introduce the random measure

$$P_{W,n} := \Sigma_{i=1}^n W_{i,n} \delta_{X_i} .$$

The idea of the (generalized) bootstrap is to approximate the distribution of $n^{1/2}(T(P_n) - T(P))$ by that of $n^{1/2}(T(P_{W,n}) - T(P_n))$ conditionally on P_n . Efron's bootstrap as defined in Efron (1979, 1982) consists in taking $\mathcal{W}_n = n^{-1}\mathcal{M}ultinomial(n,1/n,\ldots, 1/n)$. In this case $W_{i,n}$ is the frequency of X_i in the bootstrapped sample. This approach with random weights (called the resampling plan in Efron (1982) or Wu (1986)) emphasizes the double random aspect of the bootstrap. Other choices for the weights may be found in Praestgaard and Wellner (1993).

For the generalized bootstrap, Mason and Newton (1992) studied the sample mean and the empirical and quantile processes when $B = \mathbb{R}$. In particular, they proved the consistency of the bootstrap for the mean assuming the following conditions on the weights,

(1.1) $$\Sigma_{i=1}^{n} W_{i,n} = 1 ,$$

(1.2) $$\max_{1 \le k \le n} (W_{i,n} - n^{-1})^2 / \Sigma_{i=1}^{n} (W_{i,n} - n^{-1})^2 = o_p(1) \quad \text{as } n \to \infty.$$

In Haeusler, Mason and Newton (1992), the distribution $\mathcal{W}_n$ of the weights is data dependent. However, in this chapter, we will assume that

(1.3) $\mathcal{W}_n$ and the $W_{i,n}$'s are independent of the sample,

which holds for Efron's (1979), Rubin's (1981),... bootstraps.

Our purpose in this chapter is to give some asymptotic results for the generalized bootstrap of a large class of statistics in the spirit of Dudley (1990) and Giné and Zinn (1990), without restricting the X_i's to be real valued as generally required to apply the quantile-transform-empirical-process approach as in Csörgo and Mason (1989) or Einmahl and Mason (1992) (see also the volume edited by Hahn, Mason and Weiner (1991) for this approach).

Now, let us explain what kind of results we obtain. Among other things, we prove that the convergence of the generalized bootstrapped distribution to the true distribution holds in probability along the sample for first order Fréchet differentiable functionals (Theorem 3.1). Related work on Efron's bootstrap of differentiable functionals may be found in Bickel and Freedman (1981), Lohse (1984, 1987), Parr (1985), Gill (1989), Pons and Turkheim (1989). The originality of our approach is that we do not make use of continuous differentiability assumption at P . This is generally made possible by an adequate choice of the distance (maybe depending on the unknown P) which metrizes the topology on $\mathcal{P}$.

Another topic in this chapter is to investigate the convergence of the generalized bootstrap variance estimator

$$S_n^{\,2} := E_W\big(n^{1/2}(T(P_{W,n}) - T(P_n))\big)^2 / \operatorname{Var}(nW_{1,n}) ,$$

where E_W is the expectation under the distribution $\mathcal{W}_n$ of the resampling plan, conditioned on the sample (Theorem 3.2). We show that this generalized bootstrap variance estimator is generally efficient (Theorem 3.2). Our results allow us to use the generalized bootstrap variance estimator to construct confidence intervals for the parameter. The first step to obtain second order correct confidence interval is to prove an Edgeworth expansion for the studentized statistics $n^{1/2}(T(P_n) - T(P))/S_n$. This is done in Theorem 3.3 under a mild differentiability condition on $T(.)$. Again, we do not require a continuous differentiability as it is done in previous works. At this stage the choice is between bootstrapping the pivotal statistics $n^{1/2}(T(P_n) - T(P))/S_n$ to obtain automatic second order correct confidence intervals in the spirit of Hall's (1986, 1992) t-percentile method, or to invert the Edgeworth expansion after having estimated the coefficients (as done for instance in Abramovitch and Singh (1985)). Nevertheless it seems difficult and computationally expensive to bootstrap $n^{1/2}(T(P_n) - T(P))/S_n$ again, unless S_n has an explicit form. Moreover, there is no reason for S_n to behave smoothly enough at P to ensure the validity of a new bootstrap. For these reasons, we shall invert the Edgeworth expansion. In practice, the difficulty is then to estimate the coefficients, or somehow equivalently to calculate the influence function of $T(.)$ at the unknown p.m. P . This difficulty is overcome in using two different generalized bootstrap schemes (Theorem 3.5). Such a method is less computationally expensive than the double bootstrap and requires less regularity assumptions than previous methods (see for instance Beran (1984), or Tu (1992) for jackknife estimates).

This chapter is organized as follows. In section 2 we describe the functionals and the context we are interested in. In section 3 we give the main theorems and discuss the results. Section 4 is devoted to some examples and applications. Section 5 contains some simulation results and some practical results on how to generate proper weights. The proofs of the theorems are postponed until chapter IV.

2) Fréchet-differentiability and metric indexed by a class of functions

2.1) Differentiability assumption

Let us first make precise our differentiability assumptions on T . The space $\mathcal{P}$ is assumed to be endowed with a metric d . If $\mathcal{C}$ is a class of subsets of $\mathcal{P}$, we say that T is m-times $\mathcal{C}$-differentiable at P if there exists some p-linear functions D^pT_P , $0 \le p \le m$ such that for any $t \in [0,1]$, and for any $P, Q \in \mathcal{P}$,

(2.1) $$T(P + t(Q-P)) - T(P) = \Sigma_{p=1}^{m}(p!)^{-1}t^{p}D^{p}T_{P}(Q-P)^{p} + R^{(m)}(t,P,Q) ,$$

with for any $C \in \mathcal{C}$,

(2.2) $$\lim_{t \to 0} \sup_{Q \in C} |t|^{-m}|R^{(m)}(t,P,Q)| = 0$$

(see e.g. Yamamuro (1974), Reeds (1976), Flett (1980), Fernholz (1983), Gill (1989)). The special cases of interest are when

(i) $\mathcal{C}$ is the set of all bounded subsets of $\mathcal{P}$, T is m-times Fréchet-differentiable.

(ii) $\mathcal{C}$ is the set of all compact subsets of $\mathcal{P}$, T is m-times compact-differentiable or Hadamard differentiable.

(iii) $\mathcal{C}$ is the set of all subsets of $\mathcal{P}$ with only one element, T is m-times Gâteaux-differentiable.

Assuming continuous compact differentiability of T in a neighborhood of P , and using the convergence of the bootstrapped empirical process to the brownian bridge, Lohse (1984, 1987), Gill (1989) and Pons and Turkheim (1989) established the strong consistency of Efron's bootstrap. Reeds (1976) pointed out that Fréchet-differentiability easily yields central limit theorem (CLT), but very few functionals are Fréchet-differentiable with respect to the used Kolmogorov's distance, whereas they are very often Gâteaux-differentiable or tangentially Hadamard-differentiable to a separable subspace. However in that case, the CLT is not easy to prove. Continuous compact-differentiability is a kind of compromise between these two notions (see Gill (1989) for a discussion and applications).

Our approach is different in the sense that we will assume continuity and Fréchet-differentiability only at the point P , with a metric which ensures differentiability of a large class of functionals.

Before giving the distance, let us recall some facts about the representation in (2.1). By the Riesz Representation Theorem (see e.g. Rudin (1974)), the p-linear functions $D^{p}T_{p}$ admit an integral representation

$$D^{p}T_{P}(Q-P)^{p} = \int \ldots \int T^{(p)}(x_1, \ldots, x_p, P) \, d(Q-P)(x_1) \ldots d(Q-P)(x_p) ,$$

where the functions $T^{(p)}$ are not necessarily unique. In this chapter, we are essentially interested in $T^{(1)}$, $T^{(2)}$. They are uniquely defined if we assume

$$E_P\, T^{(1)}(X,P) = 0 \ ,$$

$$T^{(2)}(X,Y,P) = T^{(2)}(Y,X,P) \quad \text{(symmetry), and}$$

$$E_P\, T^{(2)}(X,y,P) = 0 \quad \text{for every } y \ .$$

Notice that $T^{(1)}$ is the influence function as defined in Hampel (1974) or Huber (1981). In Pfanzagl's (1985) terminology, $T^{(1)}(x,P)$ is no more than the first canonical gradient and $T^{(2)}(x,y,P)$ the second order canonical gradient. A very convenient way to obtain them is to calculate the following Gâteaux-derivatives :

$$T^{(1)}(x,P) = \lim_{t\to 0} t^{-1}\Big(T((1-t)P+t\delta_x) - T(P)\Big)$$

$$= \frac{d}{dt}\, T((1-t)P+t\delta_x)\Big|_{t=0} ,$$

$$T^{(2)}(x,y,P) = \lim_{\substack{t\to 0\\ t'\to 0}} (tt')^{-1}\Big(T((1-t-t')P + t\delta_x + t'\delta_y) - T(P)\Big)$$

$$= \frac{\partial^2}{\partial t\ \partial t'}\, T((1-t-t')P + t\delta_x + t'\delta_y)\Big|_{\substack{t=0\\ t'=0}}$$

To obtain results using Fréchet-differentiability, it is convenient to use a distance $d(.,.)$ which satisfies (as the Kolmogorov distance does)

$$(2.3) \qquad d(P + t(Q-P),\ P) = |t|\ d(P,Q) \ .$$

Indeed, if T is m-times Fréchet-differentiable at P for a distance d such that (2.3) holds, it is easy to check, from (2.1) and (2.2), that there exists a function $\varepsilon^{(m)}(.,P)$ such that

$$(2.4) \qquad \varepsilon^{(m)}(.,P) \text{ is continuous at } 0 \text{ with } \varepsilon^{(m)}(0,P) = 0 \ ,$$

and such that for any probability $Q \in \mathcal{P}$

$$(2.5) \qquad T(Q) - T(P) = \Sigma_{p=1}^{m}\ D^pT_P(Q-P)^p \ /\ p! + R^{(m)}(Q,P)$$

where, for short,

$$R^{(m)}(Q,P) = R^{(m)}(1,P,Q)$$

is such that

$$(2.6) \qquad |R^{(m)}(Q,P)| \leq d(P,Q)^m \, \varepsilon^{(m)}(d(P,Q),P) .$$

Unless the ball $\{ Q : d(P,Q) \leq 1 \}$ is compact, it seems difficult to establish an analogue of (2.5) and (2.6) assuming only compact differentiability.

2.2) Choice of the metric

The distance commonly used in such a context is Kolmogorov's distance (see e.g. Reeds (1976), Serfling (1980), Huber (1981), Fernholz (1983)), which satisfies (2.3). The reason for this, is that the Dvoretsky-Kiefer-Wolfowitz (1956) theorem provides a very precise control on $d(P_n,P)$ and consequently on the remainder $R^{(m)}(P,P_n)$ with (2.6). Nevertheless, this distance is not always strong enough. For instance, the mean ($T(P) = \int x \, dP(x)$) is not continuous for this distance, even if we restrict P to belong to the set of all probability measures with finite mean.

For these reasons, we are interested in a generalization of Kolmogorov's distance defined as follows : let $\mathcal{H}$ be a class of real valued measurable functions (in fact measurability is not necessary, in the sense that we can use outer measure and Hoffmann-Jørgensen's (1991) convergence), and define :

$$(2.7) \qquad d_{\mathcal{H}}(P,Q) := \sup_{h\in\mathcal{H}} \left| \int h \, d(P-Q) \right| .$$

Such a distance is used by Dudley (1987, 1990, 1992, 1994) and belongs to the class of Zolotarev's semimetric (see e.g. Rachev (1991)). For

$$\mathcal{H}_K := \{ \mathbb{I}\{.\leq y\} \ ; \ y\in\mathbb{R} \} ,$$

(2.7) defines Kolmogorov's distance in the real valued case. Taking

$$\mathcal{H}_{K,p} := \{ |.|^p \, \mathbb{I}\{.\leq y\} \ ; \ y\in\mathbb{R} \}$$

on the set of probabilities with a finite p-th moment, ensures the differentiability of $T(P) = \int |x|^p dP(x)$ for $d_{\mathcal{H}}$. Many statistical functionals are Fréchet-differentiable for $d_{\mathcal{H}}$ with a suitable $\mathcal{H}$. Moreover, condition (2.3) is satisfied by $d_{\mathcal{H}}$ for any $\mathcal{H}$. Results and references related to this distance may be found in Giné and Zinn (1991), Rachev (1991), Wellner (1991), Sheehy and Wellner (1992) and one often assumes

(2.8) $\quad 0 \le h \le 1$ for all $h \in \mathcal{H}$.

These results may be used to prove consistency of Efron's bootstrap (see for example Beran and Millar (1986), Arcones and Giné (1989, 1992)). However, condition (2.8) is extremely restrictive. The reason is that if T is Fréchet-differentiable for $d_{\mathcal{H}}$ and if (2.8) holds, then this typically forces T to be robust in the sense of Hampel (1974) i.e. that the influence function $T^{(1)}(.,P)$ is bounded (see our Proposition 2.1). In the spirit of Reeds (1976), we allow the topology on $\mathcal{P}$ (i.e. $d_{\mathcal{H}}$ or $\mathcal{H}$) to be based on the functional T itself, and even on the unknown distribution P ! This enables us to drop further conditions on differentiability or uniformity (in a neigborhood of P) usually assumed (see e.g. Bickel and Freedman (1981)).

To give some indications on how the class $\mathcal{H}$ can be chosen, we need some assumptions. For this, we assume that $\mathcal{P}$ is such that T is defined on $\mathcal{P}$, and that

(2.9) $\quad \mathcal{P}$ is convex and contains the Dirac measures.

We will assume also that our class

(2.10) $\quad \mathcal{H}$ admits an envelope H, with $E_P H(X) < \infty$

which means (see e.g. Pollard (1984) or Dudley (1984)) that for any function $h \in \mathcal{H}$, $|h| \le H$. There is no loss of generality to impose that

(2.11) $\quad$ for a constant $\gamma > 0$, and any $x \in B$, $H(x) \ge \gamma$,

since if H is an envelope of $\mathcal{H}$, so is $H \vee \gamma$.

Differentiability according to the Kolmogorov distance implies that the influence function is bounded. This is one of the main feature of robustness. Fréchet-differentiability according to a metric $d_{\mathcal{H}}$ implies that the influence function is bounded by the envelope H. A more general result is stated in the next proposition.

Proposition 2.1 : **If (2.9)-(2.11) hold and if T is m-times Fréchet-differentiable for $d_{\mathcal{H}}$, then for any $p \le m$, there exists some constants $C_p(P)$ such that**

$$|T^{(p)}(x_1, \ldots, x_p, P)| \le C_p(P)\, H(x_1) \ldots H(x_p) . \tag{2.12}$$

Remark 2.1 : It is important to notice that if for some $\rho > 0$, $E_P H(X)^{\rho} < \infty$, and if T is, say, second order Fréchet-differentiable for $d_{\mathcal{H}}$, then (2.12) implies

$$E_P\, |T^{(1)}(X_1,P)|^{\rho} < \infty , \quad E_P\, |T^{(2)}(X_1,X_2,P)|^{\rho} < \infty \ \text{ and}$$

$$E_P\, |T^{(2)}(X_1,X_1,P)|^{\rho/2} < \infty .$$

Remark 2.2 : In many situations it is easy to calculate $T^{(p)}$, at least for $p = 1$ or 2 without proving Fréchet-differentiability of T , but by just using Gâteaux-derivatives. Assume that $T^{(1)}$ and $T^{(2)}$ are known explicitly, maybe depending on P . Inequality (2.12) suggests that we may define $\mathcal{H}$ by first constructing its envelope as follows (if $B = \mathbb{R}$) :

$$H_{T,P}(x) = |T^{(1)}(x,P)| \vee \sup_{-x\le y\le x} T^{(2)}(x,y,P) \vee 1 ,$$

so that (2.12) holds with $H_{T,P}$ instead of H , and for $C_p(P) = 1$ $(p = 1,2)$. Then consider a class $\mathcal{D}$ of subsets of $\mathbb{R}$ (for example the intervals $(-\infty,x]$, $x\in\mathbb{R}$) and define

$$\mathcal{H}_{T,P} := \{\, H_{T,P}(.)\, \mathbb{I}\{.\in C\} : C\in\mathcal{D} \,\} .$$

This defines a class of functions for which, one can generally prove second order differentiability of T for $d_{\mathcal{H}}$, if $\mathcal{D}$ is sufficiently large. In some non-regular cases, a more complicated choice for $\mathcal{H}_{T,P}$ may be necessary. For instance, notice that Dudley (1992, 1994) recently proved that the inverse operator and the composition are Fréchet-differentiable with respect to a p-variation norm that admits a dual representation as a distance $d_{\mathcal{H}}$. In the real valued case, this result may be composed with our proposition to yield differentiability of more complex functionals.

2.3) Rate of convergence of the empirical process indexed by a class of functions

We now consider a general class of functions $\mathcal{H}$. In order to control the behaviour of $d_{\mathcal{H}}(P_n,P)$ and $d_{\mathcal{H}}(P_{W,n},P)$ we need some conditions on $\mathcal{H}$ (intuitively this class must not be too large) and on the weights. We strengthen (2.10) with the assumption

(2.13) $$E_P H^2(X) < \infty$$

which is a standard requirement (see e.g. Pollard (1984)). The size of $\mathcal{H}$ is given by its L_1-covering number defined as follows : for a measure Q , we define $N_1(\varepsilon,Q,\mathcal{H})$ as the smallest number of functions h_i ($1 \le i \le N_1(\varepsilon,Q,\mathcal{H})$) such that for any $h \in \mathcal{H}$,

$$\min_i E_Q|h - h_i| \le \varepsilon .$$

Very often, the class $\mathcal{H}$ is a polynomial class in the sense that there exist nonnegative constants A and D depending only on $\mathcal{H}$, such that for any $\varepsilon > 0$ and any probability Q , such that $E_Q H(X) < \infty$,

(2.14) $$N_1(\varepsilon,Q,\mathcal{H}) \le A\, (E_Q H(X)/\varepsilon)^D \vee 1$$

(see e.g. Pollard (1984), pp. 27-30). Bounds on the covering numbers of some useful classes of functions such as Lipschitzian functions can be found in Kolmogorov and Tihomirov (1959), Birman and Solomjak (1967), Dudley (1974, 1984) and Lorentz (1986).

Now, the weights are linked to the class $\mathcal{H}$ by the relation

(2.15) $$E_{W,P}(n\Sigma_{i=1}^n W_{i,n}^2 H^2(X_i))^{q/2} = O(1) \quad \text{as } n\to\infty$$

which will be assumed to hold for some special values of q . (2.15) is a Hölder type relation in the sense that it is implied (under 1.3) by

(2.16) $$E_W(n^{-1}\Sigma_{i=1}^n |nW_{i,n}|^{2\alpha})^{q/2\alpha} = O(1) \text{ , as } n\to\infty \text{ and}$$

for $\alpha^{-1} + \beta^{-1} = 1$. In nearly all practical situations, (2.16) is implied by

$$E_W|nW_{1,n}|^q = O(1) \ , \text{ as } n\to\infty.$$

And by Marcinkiewicz-Zygmund's Theorem (see e.g. Chow and Teicher (1988) Theorem 10.3.3), (2.17) is implied by

$$(2.18) \qquad E_P H^q(X_i) < \infty \ .$$

Now, we provide a rate of convergence of $P_{W,n}$ and P_n to P for a distance $d_{\mathcal{H}}$. We use $O_{W,P}$ for the order in probability according to the probability P and $\mathcal{W}_n$, O_W for the order in probability according to the probability $\mathcal{W}_n$, conditionally on $(X_1, \ldots, X_n)$ and in probability along the sample.

<u>Proposition 2.2</u> : **Assume that** (1.1)-(1.3) **and** (2.10)-(2.14) **hold, that** (2.15) **holds for some** $q > D+1$, **and that** (2.16) **holds with** $\alpha = 1$. **Then, for any** $\rho \le q$,

$$E_{P,W}\, d_{\mathcal{H}}(P_{W,n},P)^{\rho} = O(n^{-\rho/2}(\log n)^{\rho/2}) \quad \textbf{as} \quad n\longrightarrow\infty \ .$$

Therefore we have the orders in probability

$$(2.19) \qquad d_{\mathcal{H}}(P_{W,n},P) = O_P(n^{-1/2}(\log n)^{1/2}) \quad \textbf{(as} \quad n\longrightarrow\infty\textbf{)} \quad \textbf{and}$$

$$(2.20) \qquad d_{\mathcal{H}}(P_{W,n},P) = O_W(n^{-1/2}(\log n)^{1/2}) \quad \textbf{(as} \quad n\longrightarrow\infty\textbf{)}$$

Of course, the deterministic weights $W_{i,n} = n^{-1}$ may be used in Proposition 2.2 to get the order of $d_{\mathcal{H}}(P_n,P)$. Notice that, the logarithmic factors in (2.19) and (2.20) can be removed in using the recent work of Praestgaard and Wellner (1993). However, in almost all our further results (when the variance or the skewness of the distribution appears) we need to control $E_{P,W}d_{\mathcal{H}}(P_{W,n},P)^{\rho}$, for $\rho = 1, 2, 3$. This control does not seem to follow as a straight consequence of Praestgaard and Wellner (1993). Nevertheless, the technique they used may give better conditions than ours, but our conditions seem to be sufficient for usual statistical applications.

3) Consistency of the generalized bootstrapped distribution variance estimation and Edgeworth expansion.

3.1) Consistency of the generalized bootstrap

To show that the bootstrap is consistent in the sense that the asymptotic distribution of the bootstrapped statistic $T(P_{W,n})$ is the same as that of $T(P_n)$, we need to control in probability the remainder term in the expansion (2.5) when $Q = P_n$ or $P_{W,n}$. Hence, in (2.6) we will assume that for some $\delta = \delta(m) > 0$,

$$\varepsilon^{(m)}(t,P) = o(1/(-\log t)^{1+\delta}) \quad \text{as } t \to 0 \text{ , } m \geq 1 \tag{3.1}$$

which holds for example if the functional T is $(m+1)$ times Fréchet-differentiable (in this case $\varepsilon^{(m)}(t,P) = O(t)$ as $t \to 0$). We denote L (resp. Pr) the convergence in distribution (resp. in probability), and L^* (resp. Pr^*) when the convergence holds conditionally on P_n in probability along the sequence $X_1, X_2, \ldots$

If $W_{i,n}$ is square integrable, we denote

$$\sigma_{W,n}^2 := E_W(nW_{i,n}-1)^2 = \mathrm{Var}(nW_{i,n}) .$$

Theorem 3.1 : **Assume that the weights satisfy** (1.1)-(1.3), **T is first order Fréchet-differentiable (i.e.** (2.4) **and** (2.6) **hold whith** $m=1$ **) for a distance $d_{\mathcal{H}}$ such that the assumptions of Propositions 2.1 and 2.2 hold, and that** (3.1) **is valid for** $m = 1$. **Then**

$$n^{1/2}(T(P_{W,n}) - T(P_n)) \,/\, (\sigma_{W,n} S(P)) \xrightarrow{L^*} \mathcal{N}(0,1) \quad \textbf{as } n \to \infty \text{ , } \textbf{with} \tag{3.2}$$

$$S^2(P) := E_P T^{(1)}(X_1,P)^2 .$$

Remark 3.1 : In Theorem 3.1, assumption (3.1) is required only due to (2.19). For example, if instead of (2.19) we have $d_{\mathcal{H}}(P_{W,n},P_n) = O(n^{-1/2})$ (for instance for the Kolmogorov distance, or more generally a Donsker classes of functions; see Praestgaard and Wellner (1994)), then (3.1) is not necessary and first order Fréchet-differentiability of T is sufficient.

Obviously, if $T^{(1)}(.,P)$ is completely known (despite that P is unknown), one can replace $S^2(P)$ in (3.2) by

$$\hat{S}_n^2 := \mathrm{Var}_{P_n} T^{(1)}(X,P) .$$

This may happen for instance in some testing problems, when one works under the null hypothesis.

3.2) The generalized bootstrap variance estimator

Under some continuity assumptions, one can also use $S^2(P_n)$ if it exists (this is not always the case, for instance for a quantile) (see Pfanzagl (1985, 1990)). However, it seems easier, and in fact it is also less restrictive on T, to use a bootstrapped variance estimator, defined as the variance of the distribution of $T(P_{W,n}) - T(P_n)$ given P_n, and which we denote

$$(3.3) \qquad S_n^2 := E_W(n^{1/2}(T(P_{W,n})-T(P_n)))^2/\sigma_{W,n}^{\ 2} .$$

Since $S_n^{\ 2}$ is obviously an estimator for the asymptotic variance of $T(P_n)$ we may expect that $S_n^{\ 2}$ behaves nicely. However, notice that the weighted probability $P_{W,n}$ may be quite far (for the distance $d_{\mathcal{H}}$) from P with a small probability. More precisely we mean, that when we take the expectation in (3.3) w.r.t. to the probability of the weights, the measure $P_{W,n}$ is varying over almost all the set $\mathcal{P}$; consequently, if we want to obtain some results on $S_n^{\ 2}$, we must control the behaviour of the functional T over the whole space $\mathcal{P}$. If T is twice Fréchet-differentiable, the most convenient way to get this control is to assume that

$$(3.4) \qquad \varepsilon^{(2)}(t,P) \le K_\infty(P)\, t^{\mu} \quad \text{as} \quad t \to \infty$$

for some nonnegative constants $K_\infty(P)$ and $\mu = \mu(P)$, and $\varepsilon^{(2)}(t,P)$ defined as in (2.4)-(2.6). The real meaning of (3.4) has nothing to do with differentiability of T, but bounds the variation of $T(Q) - T(P)$ by a function of $d_{\mathcal{H}}(P,Q)$ when Q is far from P.

Now, notice that $P_{W,n}$ is far from P_n typically if some weights $W_{i,n}$ are too large or too small compared to $1/n$. Therefore, it is not surprising that the control on T given in (3.4) is linked with some properties of the weights, and more precisely with the fact that they admit some moments. So, we will assume conditions like

(3.5) $$\limsup_{n\to\infty} E_W|nW_{i,n}-1|^r < \infty$$

where r will be related to μ in (3.4) with the help of Proposition 2.2.

Then, for technical reasons, we strengthen a little the equivalent of (3.1) at the second order by assuming that

(3.6) $$\varepsilon^{(2)}(t,P) \le K_0(P)\, t^{\lambda}\ , \text{ as } t\to 0,$$

for some positive constants $K_0(P)$, $\lambda = \lambda(P)$. If T is third-order Fréchet-differentiable, one may take $\lambda = 1$ or any $\lambda \le 1$.

If (3.4) holds for some μ, it also holds if we increase μ and on the other hand, if (3.6) holds for some positive λ, it also holds if we decrease λ. Since all what we really need is the existence of λ (arbitrarily small), there is no loss of generality in assuming

(3.7) $$\lambda \le \mu\ .$$

The following theorem shows that the estimator S_n^2 is not only convergent but also efficient in many situations.

<u>Theorem 3.2</u> : **Assume that**

- **T is twice Fréchet-differentiable for a distance $d_{\mathcal{H}}$,**
- **$\mathcal{H}$ admits an envelope H with $E_P H(X)^4 < \infty$,**
- **(3.4), (3.6) and (3.7) hold,**
- **Proposition 2.1 holds for $m = 2$,**
- **Proposition 2.2. holds for some $q \ge 4(2+\mu)/(1+\lambda)$,**
- **(3.5) holds for $r = 4$ and $\liminf_{n\to\infty} \sigma_n^2 > 0$.**

Then,

i) S_n^2 converges in probability to $S^2(P)$, and

ii) S_n^2 admits a stochastic expansion

(3.8) $$S_n^2 - S^2(P) = n^{-1}\Sigma_{i=1}^n S^{(1)}(X_i,P) + o_p(n^{-1/2}) \quad \text{with}$$

(3.9) $$S^{(1)}(.,P) := T^{(1)}(.,P)^2 - S^2(P) + 2\int T^{(1)}(x,P)\, T^{(2)}(x,.,P)\, dP(x)\ .$$

iii) If $S^2(P)$ is first order Fréchet-differentiable for some $d_{\mathcal{H}^*}$, for some

$\mathcal{H}^*$ maybe different from $\mathcal{H}$, then

$$(3.10)\qquad I(P) := E_P\left(T^{(1)}(X,P)^2 - s^2(P) + 2\int T^{(1)}(y,P)T^{(2)}(X,y,P)\, dP(y)\right)^2$$

is the semiparametric efficiency bound of $s^2(P)$. Hence, if ii) holds, S_n^2 is an efficient estimator of $s^2(P)$.

<u>Remark 3.2</u> : It is important in Theorem 3.2 and in many results in this work to assume (1.1). Suppose, for instance, that we change (1.1) in $E_W W_{i,n} = n^{-1}$. Then $P_{W,n}$ may not be a probability measure since its total mass becomes

$$\overline{W}_n = \Sigma_{i=1}^n W_{i,n} .$$

By writing,

$$T(P_{W,n}) - T(P_n) = T(P_{W,n}) - T(\overline{W}_n P) + T(\overline{W}_n P) - T(P_n) .$$

we see that we must define T on a set of measures with finite total mass (this is not always clear how to do it, for instance if T is a quantile). Then, if we assume differentiability of T at $\overline{W}_n P$, we shall need some uniformity of this differentiability in a neighborhood of P .

Whereas if we use the decomposition

$$T(P_{W,n}) - T(P_n) = T(P_{W,n}) - T(P) + T(P) - T(P_n) ,$$

the fact that $P_{W,n}$ is not normalized to a unit mass implies that we cannot normalize the first gradient in assuming $E_P T^{(1)}(X,P) = 0$ when we apply the differentiability assumption on $T(P_{W,n}) - T(P)$.

Therefore, when dealing with a general functional, we see that assumption (1.1) can hardly be removed.

Another proposal (see Haeusler, Mason and Newton (1992) and Hall and Mammen (1992)) is to use

$$T(P_{W,n}) - T(\overline{W}_n P_n) .$$

One can easily see that the same problems occur and that this kind of bootstrap requires in fact some uniformity assumptions on the differentiability of T .

Typically, Efron's bootstrap and the Bayesian bootstrap satisfy the conditions required on the weights in Theorem 3.2, so that they generally lead to efficient variance estimators of the asymptotic variance. This result is particularly important to construct confidence intervals for $T(P)$ based on the asymptotical pivotal statistic $n^{1/2}(T(P_n) - T(P))/S_n$. Akahira and Takeuchi (1991) obtained a result on the efficiency of Efron's bootstrap variance estimator in the particular case when $T^{(2)}(x,y,P) = 0$, so that they have $I(P) = E_P(T^{(1)}(X,P)^2 - S^2(P))^2$.

3.3) Edgeworth expansion of the studentized functional

As mentioned in our introduction, Beran (1987) and Hall (1986, 1988b, 1990) suggest that before bootstrapping one should make the quantity $n^{1/2}(T(P_n) - T(P))$ asymptotically pivotal in order to obtain more accurate confidence intervals. Obviously, Theorem 3.2 provides a simple way to do it. Hence, our next goal is to derive an Edgeworth expansion for the distribution of $n^{1/2}(T(P_n) - T(P))/S_n$. Then, we will invert it and obtain second order correct confidence intervals in estimating the coefficients of the Edgeworth expansion with a suitable choice of the weights. This approach requires weaker assumptions than the one which consists in bootstrapping the studentized statistics and has the same accuracy.

In the following theorem, we consider Edgeworth expansion for

$$H_n(x,P) := P\{ n^{1/2}(T(P_n) - T(P))/S(P) \le x \} \quad \text{and}$$

$$K_n(x,P) := P\{ n^{1/2}(T(P_n) - T(P))/S_n \le x \} .$$

For $H_n(x,P)$ the Edgeworth expansion may be deduced from Götze (1979, 1984). We will need the usual Cramér condition which becomes here

$$\text{(3.11)} \qquad \limsup_{t\to\infty} |E_P \exp(itT^{(1)}(x,P))| < 1 \quad \text{and}$$

$$\text{(3.12)} \qquad \text{for any } t > 0 \text{ , } |E_P \exp(itT^{(1)}(x,P))| < 1 \text{ .}$$

We also introduce the following coefficients :

$$K_{1,1}(P) := E_P T^{(2)}(X_1,X_1,P) \,/S(P)\ ,$$

$$K_{1,3}(P) := S(P)^{-3}\ E_P\Big(T^{(1)}(X_1,P)^3 +$$

$$3T^{(1)}(X_1,P)T^{(1)}(X_2,P)T^{(2)}(X_1,X_2,P)\Big)$$

(3.13) $$K'_{1,1}(P) := E_P T^{(2)}(X_1,X_1,P)/S(P) - E_P T^{(1)}(X_1,P)^3/S(P)^3$$

$$- 2\ E_P T^{(1)}(X_1,P)T^{(1)}(X_2,P)T^{(2)}(X_1,X_2,P)/S(P)^3$$

(3.14) $$K'_{1,3}(P) := -2E_P T^{(1)}(X_1,P)^3/(2S(P)^3)$$

$$- 3\ E_P T^{(1)}(X_1,P)T^{(1)}(X_2,P)T^{(2)}(X_1,X_2,P)/S(P)^3\ .$$

Finally, we denote by $\Phi(x)$ the cumulative distribution function of the standard normal distribution, and by $\phi(x) := (2\pi)^{-1/2}\exp(-x^2/2)$ its density function.

Theorem 3.3 **a) Assume that**

(3.15) . T **is twice Fréchet-differentiable at** P **for a metric** $d_{\mathcal{H}}$, **and** (3.1) **holds for** $m = 2$,

(3.16) . (3.11) **and** (3.12) **hold,**

(3.17) . **assumptions of Proposition 2.1 hold for** $m = 2$,

(3.18) . (2.14) **holds**

(3.19) . (2.18) **holds for** $q > 2(D+2)$

then the following Edgeworth expansion is valid :

(3.20) $$\sup_{-\infty<x<\infty} \Big| H_n(x,P) - \Phi(x) + n^{-1/2}\ 2^{-1}\ K_{1,1}(P)\ \phi(x)$$

$$+ n^{-1/2}\ 6^{-1}\ K_{1,3}(P)\ (x^2-1)\phi(x)\ \Big| = o(n^{-1/2}) \quad \text{as} \quad n\to\infty\ .$$

b) Assume that

(3.21) . **the assumptions of Theorem 3.2 hold,**

(3.22) . $0 < \liminf_{n\to\infty} \sigma^2_{W,n} \le \limsup_{n\to\infty} \sigma^2_{W,n} < \infty$ **and** (3.5) **holds for** $r = 4$,

(3.23) . (2.18) **holds for some** $q > 6 \vee (D+1)$

(3.24) . **Proposition 2.2 holds for some** $q > 3(2+\mu)$.

Then K_n admits the Edgeworth expansion

(3.25) $$\sup_{-\infty<x<\infty} \Big| K_n(x,P) - \Phi(x) + n^{-1/2}\, 2^{-1}\, K'_{1,1}(P)\, \phi(x)$$

$$+ n^{-1/2}\, 6^{-1}\, K'_{1,3}(P)\, (x^2-1)\phi(x) \Big| = o(n^{-1/2}) \quad \textbf{as} \quad n\to\infty .$$

Remark 3.3 : The meaning of (3.22) is clear for Efron's bootstrap : it simply says that the resampling size is of the same order as the sample size. For the generalized bootstrap, it says that each term δ_{X_i} of P_n keeps a contribution in the bootstrapped probability $P_{W,n}$ as $n\to\infty$, and that this contribution has a random character and does not degenerate to something deterministic like $1/n$.

Remark 3.4 : Proposition 2.1 ensures that under (3.19) or (3.23), all the coefficients $K_{1,1}(P)$, $K_{1,3}(P)$, $K'_{1,1}(P)$ and $K'_{1,3}(P)$ are finite.

3.4) Inverting the Edgeworth expansion to construct confidence intervals

Our next step is to provide estimates for $K_{1,1}(P)$, $K_{1,3}(P)$ and invert the Edgeworth expansion given by (3.20). For this, we define for any integer p for which the expression has a meaning,

$$\mu_{p,n} := n^{1/2}\, E_W(n^{1/2}(T(P_{W,n})-T(P_n)))^p .$$

Then, let

$$\beta_{W,n} := E_W(nW_{1,n}-1)^3/\sigma_{W,n}^{\ 3}$$

$$K_{1,n} := 2\mu_{1,n} \,/\, (S_n\sigma_{W,n}^{\ 2}) \quad \text{and}$$

$$K_{3,n} := \Big(\mu_{3,n} - 3\, S_n^{\ 2}\, E_W\big((nW_{1,n}-1)^2(nW_{2,n}-1)^2\big)\, \mu_{1,n}/\sigma_{W,n}^{\ 2}\Big)$$

$$/\big(S_n^{\ 3}\, \beta_{W,n}\sigma_{W,n}^{\ 3}\big) .$$

In the following theorem [x] denotes the smallest even integer strictly greater than x.

Theorem 3.4 : **Assume that**

(3.26) . **the hypotheses of Theorem 3.3.a) hold,**

(3.27) . (3.5) **holds for** $r = 6$

(3.28) . $E_W(nW_{1,n}-1)^3 = E_W((nW_{1,n}-1)^2(nW_{2,n}-1)^2) + o(n^{-1/2})$ **as** $n\to\infty$,

(3.29) . (2.18) **holds for some** $q > 6$,

(3.30) . **the hypotheses of Theorem 3.2 hold for some** $q > ([3/\lambda] \vee 4)(2+\mu)$,

then, we have

(3.31) $$\lim_{n\to\infty} K_{1,n} = K_{1,1}(P) \quad \text{in probability} ,$$

(3.32) $$\lim_{n\to\infty} K_{3,n} = K_{1,3}(P) \quad \text{in probability, and}$$

for any compact set $C \subset \mathbb{R}$,

(3.33) $$\sup_{x\in C} \left| P\{n^{1/2}(T(P_n)-T(P))/S(P) \le x - n^{-1/2}2^{-1}K_{1,n} - n^{-1/2}6^{-1}K_{3,n}(x^2-1) \} - \Phi(x) \right| = o(n^{-1/2}) .$$

Remark 3.5 : The condition (3.30) with $q > ([3/\lambda] \vee 4)(2+\mu)$ which is used to obtain (3.33), has the following meaning : if $\lambda = 1$ (i.e. typically if T is third order differentiable), we see that we require $E_P\, H(X)^{4(2+\mu)}$ which ensures that $K_{3,n}$ behaves nicely. Now if λ must be chosen smaller than 1, our functional is twice differentiable but not third order differentiable. In some sense the smaller $\lambda > 0$ is, the higher the singularity of T at P is. Thus it is not surprising that for small λ we must add more regularity conditions on the moments of H and $W_{i,n}$ to invert the Edgeworth expansion.

Remark 3.6 : Condition (3.28) is satisfied for Efron's bootstrap with a resampling size equal to the sample size. It does not hold for the Bayesian bootstrap. However, if we take the weights as

(3.34) $$W_{i,n} = Y_i \,/\, (Y_1 + \dots + Y_n)$$

where $(Y_i)_{i\ge1}$ is an i.i.d. sequence of Gamma(2) distributed r.v. with common density function $\gamma_2(x) := xe^{-x}$, then (3.28) holds. In fact, for non-negative weights defined as in (3.34), (3.28) often means that

(3.35) $$m\, E(Y_1-m)^3 = (E(Y_1-m)^2)^2 , \quad \text{where}$$

(3.36) $$m := EY_1 \ .$$

(see Appendix 1 where the moments of the $W_{i,n}$ in (3.34) are related to those of Y_i).

For example, log-normal r.v.'s Y_i such that

$$\mathrm{Var} \log Y_i = 1/\log 3$$

satisfy (3.35), (3.36) and may be used in (3.34) so that (3.28) holds.

Our final goal is to estimate $K'_{1,1}(P)$ and $K'_{1,3}(P)$. An unsatisfactory and computer expensive solution is to bootstrap the statistic $n^{1/2}(T(P_n)-T(P))/S_n$ as indicated in the previous section.

However, notice that $K'_{1,1}(P)$ and $K'_{1,3}(P)$ depend only on

$$S(P)^2, \quad E_P T^{(2)}(X,X,P) \quad \text{and}$$

(3.37) $$E_P T^{(1)}(X,P)^3 \ , \quad E_P T^{(1)}(X_1,P) T^{(1)}(X_2,P) T^{(2)}(X_1,X_2,P) \ .$$

In fact, we can provide estimators for all these quantities. The expectation

$$a := E_P T^{(1)}(X,P)^3$$

seems to be a key quantity. It also appears in the bias-corrected-accelerated confidence interval (BCAI) introduced by Efron (1987) (see also DiCiccio and Romano (1988)). Indeed, following Hall (1988b), if T is twice differentiable at P and if $S(P)$ is also differentiable, under some additional regularity conditions, it may be proved that the so-called accelerator (see Efron (1987) for a review) is given by $n^{-1/2} \ a/(6S^3(P))$ (see also Konishi (1991)). Efron (1987) established the validity of the BCAI in a parametric context for exponential families, and proposed an estimator of this quantity, in a nonparametric context, based on the empirical version of the influence function. Beran (1984), Tu (1992) also proposed to estimate this quantity by jackknife methods.

Since we will use various weights, we index our estimators by the resampling scheme, and therefore will denote

$$\mu_{p,n}(W) := n^{1/2} \ E_W(n^{1/2}(T(P_{W,n})-T(P_n)))^p$$

to indicate that we use the weights $W_n := (W_{1,n}, \ldots, W_{n,n})$, and similarly $\mu_{p,n}(I)$ for some weights $I_n = (I_{1,n}, \ldots, I_{n,n})$. In fact, S_n^2 and $K_{1,n}$ are already estimators for $S^2(P)$ and $E_P T^{(2)}(X,X,P)$.

Now, define for two independent weights I, J

$$a_n(I,J) := \big(\mu_{3,n}(I) - \mu_{3,n}(J)\big) \,/\, E\big((nI_{1,n}-1)^3 - (nJ_{1,n}-1)^3\big) .$$

We also have

$$(S(P))^3 K_{1,3}(P) - a)/3 = E_P\big(T^{(1)}(X_1,P)\, T^{(1)}(X_2,P)\, T^{(2)}(X_1,X_2,P)\big)$$

Therefore by plugging estimators for $K_{1,3}(P)$ and a into this expression, we obtain a natural estimator for the second quantity in (3.37). Now, we write $K'_{1,n}$ and $K'_{3,n}$ when we replace all the expectations in (3.13) and (3.14) by their estimators (which may depend on various weights). The following theorem shows that everything works and provides second order correct confidence intervals for $T(P)$.

Theorem 3.5 : **Assume that the weights** I, J, **the functional** T **and the class of functions** $\mathcal{H}$ **satisfy** (3.26), (3.27), (3.29), (3.30), **and**

(3.38) $$\lim_{n\to\infty} E_I(nI_{1,n}-1)^2(nI_{2,n}-1)^2 - (nJ_{1,n}-1)^2(nJ_{2,n}-1)^2 = 0 \quad \text{and}$$

0 **is not a limit point of the sequence**

$$\big(\, E_I(nI_{1,n}-1)^3 - E_J(nJ_{1,n}-1)^3 \,\big)_{n\geq 1} .$$

Then, we have

(3.39) $$\lim_{n\to\infty} a_n(I,J) = a \quad \textbf{in probability,}$$

and for any compact set $C \subset \mathbb{R}$,

(3.40) $$\sup_{x\in C} \Big| P\{\, n^{1/2}(T(P_n)-T(P))/S_n \leq x - n^{-1/2}2^{-1}K'_{1\,n} - n^{-1/2}6^{-1}K'_{3,n}(x^2-1) \,\} - \Phi(x) \Big| = o(n^{-1/2}) .$$

Notice that (3.38) means practically that the weights I and J have the same variance. Moreover, since (3.38) is not compatible with (3.28), we may obtain non-consistent estimators of $K_{i,j}(P)$ (at least for one of the weights).

Remark 3.7 : Notice that if one wishes to construct a bilateral confidence interval for the functional $T(P)$, one needs to establish an Edgeworth expansion for $n^{1/2}|T(P_n)-T(P))|/S_n$ up to $o(n^{-1})$ because the term of order $n^{-1/2}$ vanishes in the Edgeworth expansion. This will be done in chapter II (see (II.5.9)). The same trick of using different resampling plans to obtain estimators of the coefficients may be used in that case, yielding third order correct confidence intervals.

4) Applications

4.1) Example 1 : the mean of a function of X

Consider the functional $T(P) = E_P g(X)$, where X is a B-valued r.v., and g is a continuous function from the locally compact Banach space B into the real line $\mathbb{R}$. This function is defined on the space $\mathcal{P}$ of all probability measures such that $T(P)$ is finite.

Clearly $T(P)$ is linear in P. Consequently, if T is differentiable at P and we have $T^{(1)}(x,P) = g(x) - E_P g(X)$. Therefore, all that we have left to check is the differentiability of T for a suitable topology on $\mathcal{P}$.

Let $\mathcal{C}$ be a polynomial class of set which contains the set $\mathcal{B}$ and let

$$\mathcal{H} = \{ \ (|g(.)| \vee 1) \ \mathbb{1}_{x \in C} \ , \ C \in \mathcal{C} \ \} \ .$$

Then obviously

$$|T(Q) - T(P)| = \left| \int_{\mathcal{B}} g(x) d(P-Q)(x) \right| \le 2 \ d_{\mathcal{H}}(P,Q)$$

so that $T(.)$ is continuous at P for $d_{\mathcal{H}}$, and differentiable since linear, and all our results may be applied.

4.2) Example 2 : M-estimators

Arcones and Giné (1992) give conditions to obtain Fréchet-differentiability of functionals associated to M-estimators, with respect to a metric $d_{\mathcal{H}}$. They provide an explicit formula for the class $\mathcal{H}$. Their results may be carried over in our framework, leading to improved confidence intervals for M-estimators.

4.3) Example 3 : The probability of being censored

In this example, our observations X_i are $\mathbb{R}^2$-valued. It is more convenient to work with distribution functions instead of probability measures. Consider a bivariate distribution function F with marginal distributions F_1 and F_2. The functional $T(F) = \int F_1 \, dF_2$ plays an important role in many fields such as survival analysis, product integration or Markov process (see Gill (1989)) and may be interpreted, in some cases, as the probability of being censored. This functional is Hadamard (or compact-differentiable) but not Fréchet-differentiable with respect to the sup-norm (see Wellner (1991)). Moreover Gill (1989) shows that it is not continuously Hadamard-differentiable so that his bootstrap results for continuous compact-differentiable functionals do not apply.

We shall use that this functional is quadratic We have for any G with marginal laws (G_1,G_2) and any K with marginal laws (K_1,K_2) :

$$T(G) - T(F) = \int F_1 \, d(G_2 - F_2) + \int G_2 \, d(G_1 - F_1)$$

$$+ 2^{-1}\Big(2 \int (G_2 - F_2) \, d(G_1 - F_1)\Big)$$

where the function

$$G \longrightarrow \int F_1 \, d\, G_2 + \int G_2 \, d\, G_1$$

is clearly linear, and the function

$$(G,K) \longrightarrow 2 \int G_2 \, d\, K_1$$

is bilinear. A straightforward calculation shows that the canonical gradients are given by

$$T^{(1)}((x_1,x_2),F) = F_1(x_2) - F_2(x_1) + 1 - 2\int F_1\, dF_2 \quad \text{and}$$

$$T^{(2)}((x_1,x_2),(y_1,y_2),F) = \mathbb{1}_{y_1<x_2} + \mathbb{1}_{x_1<y_2} + F_2(x_1) + F_2(y_1)$$

$$- F_1(x_2) - F_1(y_2) + 2\int F_1\, dF_2 - 2$$

Since T(F) is quadratic we have an expansion similar to (2.5) with m=2 but without any remainder. Since the moments $\int F_1^p\, dF_2$ and $\int F_1^p\, dF_2$ are finite for any p it suffices to choose some weights satisfying (1.1)-(1.3) and $E|nW_{i,n}-1|^8 < \infty$ (which is the case for Efron's (1979) bootstrap) and the results of section 3 apply.

To conclude, notice that Dudley (1992) uses p-variation metric (for p<2) to prove that the functional T(F) is Fréchet-differentiable. These metrics may also be written as a Zolotarev type metric $d_{\mathcal{H}}$ (Rachev (1991)).

4.4) Example 4 : Multivariate V-statistics

The preceeding example suggests a more general approach for multivariate V-statistics. Assume that our observations X_i are $\mathbb{R}^d$-valued, and consider

$$T(P) = \int \omega(u_1,u_2,\ldots,u_p)\, dP(u_1)..dP(u_p)$$

where ω is a function from $(\mathbb{R}^k)^p$ into $\mathbb{R}$. We have the Hoeffding decomposition (see for instance Serfling (1980, p.215)) :

$$T(Q) - T(P)$$

$$= \Sigma_{i=1}^p \int \omega(u_1,u_2,\ldots,u_i,u_{i+1},..,u_p)\, \Pi_{j=1}^i d(Q-P)(u_j)\, \Pi_{j=i+1}^p dP(u_j)\ .$$

For $1 \le i \le p$, consider the projections,

$$\pi^{(i)}(x_1,\ldots,x_i,P) = \int \omega(u_1,u_2,,u_i,u_{i+1},..,u_p)\, \Pi_{j=i+1}^p\, dP(u_j)\ .$$

By construction, T admits a representation of type (2.5) with no remainder.

It is easy using Theorems 1-4 to find conditions on the moments of ω and on the weights for the generalized bootstrap of T to work.

5) Some simulation results

The following simulations illustrate the results obtained in Theorems 3.2, 3.4 and 3.5, respectively on the mean and the functional $T(F) = \int F_1 dF_2$ (see 4.2). We consider three types of weights.

The weights $W^{(1)} := \{ W^{(1)}_{i,n} , 1 \le i \le n \}$ are multinomial, corresponding to Efron's Bootstrap.

The weights $W^{(2)} := \{ W^{(2)}_{i,n} , 1 \le i \le n \}$ are given by

$$W^{(2)}_{i,n} := Y^{(2)}_{i,n} / \Sigma^n_{i=1} Y^{(2)}_{i,n} , \tag{5.1}$$

where the $Y^{(2)}_{i,n}$ are i.i.d. random variables with distribution a mixture of two Bernoulli distributions

$$\gamma\ B(1,p_1) + (1-\gamma)B(1,p_2) . \tag{5.2}$$

The parameters $\overline{m}$, γ , p_1 , p_2 will later be chosen to generate the $W^{(2)}_{i,n}$ with some given moments.

The weights $W^{(3)} := \{ W^{(3)}_{i,n} , 1 \le i \le n \}$ are also of the form (5.1) but $Y^{(3)}_{i,n}$ is given by

$$Y^{(3)}_{i,n} = \overline{m} + \sigma\{ N_{i,n} + 6\beta(N^2_{i,n}-1) + 24\gamma(N^3_{i,n} - 3N_{i,n}) \} \tag{5.3}$$

where $\{ N_{i,n}, 1 \le i \le n \}$ is a triangular array of independent standard gaussian r.v.'s.

We first begin with some considerations about the simulation of the exchangeable weights with some given moments. See Devroye (1986) for more details on the generation of i.i.d. weights with given moments.

In the following examples we only need to adjust the three first moments of the weights. Denote

$$m_{y,n}^{(j)} := E\, Y_{1,n}^{(j)} ,$$

$$\sigma_{y,n}^{(j)2} := \text{Var } Y_{1,n}^{(j)} \quad \text{and}$$

$$\beta_{y,n}^{(j)} := \text{Skew}(Y_{1,n}^{(j)}) .$$

We may always choose $m_{y,n}^{(j)} = 1$. It is easy to generate $W^{(1)}$ since in that case we only need to resample with replacement from the original sample. For $W^{(2)}$ and $W^{(3)}$ we have by a straightforward calculation (see Appendix 1 , Proposition A.1.3),

$$m_{W,n}^{(j)} := E_W\, n\, W_{i,n}^{(j)} = 1,$$

$$\sigma_{W,n}^{(j)^2} = \text{Var}(n\, W_{i,n}^{(j)}) = \sigma_{y,n}^{(j)^2} + n^{-1}(3\,\sigma_{y,n}^{(j)^4} - \sigma_{y,n}^{(j)^2} - 2\sigma_{y,n}^{(j)^3}\beta_{y,n}^{(j)})$$

$$+ O(n^{-3/2}) ,$$

$$\beta_{W,n}^{(j)} = \text{Skew}(n\, W_{i,n}^{(j)}) = \beta_{y,n}^{(j)} + O(n^{-1/2}), \quad \text{for } 2 \le j \le 3 ,$$

where $\text{Skew}(Y) := E(Y-EY)^3/\text{Var}(Y)^{3/2}$ is the standardized skewness of Y and $\text{Kur}(Y) := E(Y-EY)^4/\text{Var}(Y)^2$ the standardized kurtosis.

Thus it will suffice to generate i.i.d. r.v.'s $Y_{i,n}$ with mean, variance and skewness

$$m_{y,n}^{(j)} = 1$$

$$\sigma_{y,n}^{(j)^2} = \sigma_{W,n}^{(j)^2} - n^{-1}(3\,\sigma_{W,n}^{(j)^4} - \sigma_{W,n}^{(j)^2} - 2\sigma_{W,n}^{(j)^3}\beta_{W,n}^{(j)}) + O(n^{-3/2})$$

$$\beta_{y,n}^{(j)} = \beta_{W,n}^{(j)} + O(n^{-1/2})$$

where $m_{W,n}^{(j)}$, $\sigma_{W,n}^{(j)^2}$, $\beta_{W,n}^{(j)}$ will be prescribed. It seems better in the simulations to take into account the first order term although the induced error on the distributions is at least $O(n^{-1})$. This will be crucial when we shall study the third order adequacy of the generalized Bootstrap distribution in chapter II.

The mean, the variance and the skewness of $Y_{i,n}^{(2)}$, j = 2, 3 are functions respectively of the parameters $\overline{m}$, γ, p_1, p_2. A straightforward calculation allows one to calculate the moments of $Y_{i,n}^{(2)}$ and it suffices to adjust $\overline{m}$, γ, p_1 and p_2 to obtain the prescribed moments (see Devroye (1986)). In the simulations we use Devroye's (1986, p. 690) algorithm. It is easy to adapt his method to obtain a non lattice unimodal distribution by multiplying the preeceding $Y_{i,n}^{(2)}$ by an uniform random variable.

The construction of the weights $W^{(3)}$ is subject to more restrictions on the moments which can effectively be prescribed (see Devroye (1986)). However they can easily be generated in many applications, at least if the sample size is reasonable as we shall see later. A straightforward calculation shows that the standardized moments of the $Y_{i,n}^{(3)}$ are

$$E\, Y_{i,n}^{(3)} = \overline{m} = 1,$$

$$\mathrm{Var}(Y_{i,n}^{(3)}) = \sigma^2\,(1 + 2\,\beta^2 + 6\gamma^2)\,,$$

$$\mathrm{Skew}(Y_{i,n}^{(3)}) = \beta\,(1 + 4\beta^2/3 + 18\gamma^2 + 6\gamma)/(1 + 2\beta^2 + 6\gamma^2)\,,$$

$$\mathrm{Kur}(Y_{i,n}^{(3)}) = \gamma + (2\beta^4 + 135\gamma^4 + 42\gamma^3 + 2\beta^2 + 9\gamma^2 + 20\beta^2\gamma$$

$$+ 90\beta^2\gamma^2 - 4\beta^4\gamma - 36\gamma^5 - 24\beta^2\gamma^3)/\,(1 + 2\beta^2 + 6\gamma^2)\,.$$

Thus generating the $Y_{i,n}^{(3)}$ with some fixed mean, variance, skewness and kurtosis amounts to solving a nonlinear system with four parameters. It is done numerically by iterations.

Table 1 gives the values of the desired and simulated moments of $W^{(1)}$, $W^{(2)}$ and $W^{(3)}$ over nboot replications. To illustrate Theorem 3.5 we need two different types of weights in the sense that their skewness are different (see condition (3.38)). This explains the value of $\beta_{W,n}$ for $W^{(3)}$.

	$\sigma^2_{W,n}$	$\beta_{W,n}$	$\sigma^2_{W,n}$ sim	$\beta_{W,n}$ sim
$W^{(1)}$(Efron)	1.01	1.00	1.017	1.02
$W^{(2)}$(Bin)	1.00	1.00	1.016	1.16
$W^{(3)}$(Gauss)	1.00	-0.50	1.019	-0.456

(sim = simulated)

Table 1: True and empirical moments of the weights over 500 replications

It seems from these results that it is quite difficult to generate exchangeable weights with some fixed moments. That may be a serious drawback when the initial sample size is small. However, we shall see that for moderate sample sizes, the generalized Bootstrap gives very accurate results.

a) Simulation 1 : $T(P) = E_P X$.

The parameters of the simulations are the following. The r.v.'s $X_1, \ldots, X_n$ are independent and exponentially distributed with mean 1 . The sample size is $n = 100$. The generalized Bootstrap distribution is approximated by Monte-Carlo simulations. The number of bootstrap resamplings is equal to nboot = 500. We generate nboot sets of weights $\{ W^{(j)}_{i,n} , i=1,..n\}$, $1 \le j \le 3$. The results in the tables are means over nrep = 200 repetitions of the procedure. All the results have been computed on double precision using the software Gauss on a 486-PC.

Table 2 illustrates Theorems 3.2 and 3.4. It gives the true value of the quantities $S^2(P)$ and $K_{1,3}(P)$ as well as the empirical estimation and the Bootstrap estimates. Only $W^{(1)}$ and $W^{(2)}$ satisfy condition (2.28). This explains the bad result for $K_{1,3}(P)$ with $W^{(3)}$. The standard deviations of the estimates over the 200 repetitions of the procedure are indicated in parentheses. The empirical Bootstrap variance estimator seems to give good results. In average, the results are also good for the kurtosis, but the standard deviation is quite large, even for the empirical value.

	$S^2(P)$	$K_{1,3}(P)$
True	1.000	2.000
Empiric.	0.989 (0.267)	1.879 (1.120)
Est. $W^{(1)}$	0.985 (0.267)	1.930 (2.943)
Est. $W^{(2)}$	1.008 (0.269)	2.029 (2.226)
Est. $W^{(3)}$	1.004 (0.275	0.892 (1.784)

Table 2 : Empirical and Bootstrap estimators of $S^2(P)$ and $K_{1,3}(P)$, n = 100, nboot = 500, mean and standard deviation over 200 replications

Figure 1 compares the left tail of the different Bootstrap distributions, of the true and asymptotic ones. It is clear in this example that, when the weights are chosen adequately, the weighted Bootstrap can give a more accurate approximation to the true distribution than Efron's Bootstrap or the asymptotic distribution do. This was not the case for $W^{(3)}$ which gives a misleading approximation because of its inadequate skewness.

Table 3 illustrates the properties of Theorem 3.6. The bootstrap estimators of $K'_{1,1}(P)$, $K'_{1,3}(P)$ are computed using two different resampling plans (with two different kurtosis). Here using respectively $W^{(1)}$ and $W^{(3)}$, $W^{(2)}$ and $W^{(3)}$ we are able to compare two different Bootstrap estimators. One more time, in average the generalized Bootstrap works very well and provides very accurate approximation of the true distribution. However the standard deviations are large. This seems rather due to the small size of the sample rather than to the Bootstrap resampling procedure itself. Notice that the empirical estimator has a large standard deviation too. The Bootstrap estimators of $K'_{1,1}(P)$ are based on a primary estimation of $K'_{1,3}(P)$ whereas

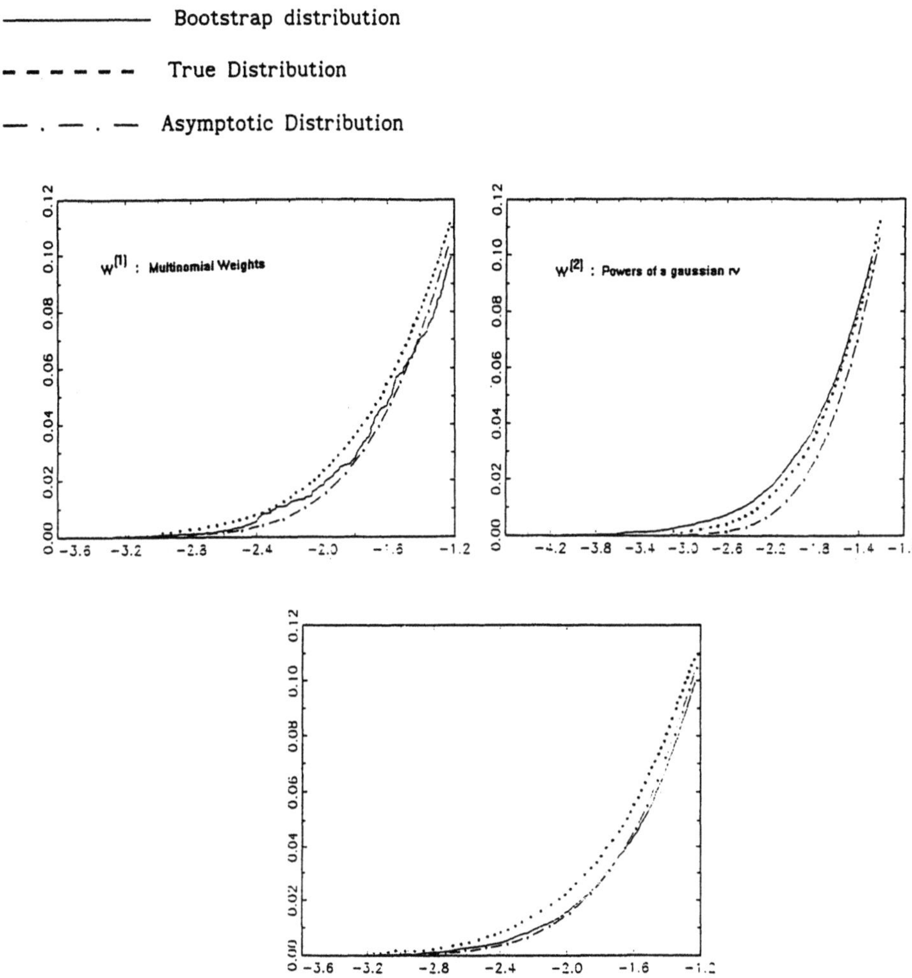

Figure 1 : Comparison of the Bootstrap Distributions with the true and asymptotic distributions

the empirical estimator is calculated directly : this explains why the Bootstrap estimators perform badly in that case, in comparison to the empirical estimator.

	a	$K_{1,1}(P)$	$K'_{1,3}(P)$	$K'_{1,1}(P)$
True	2.00	0.00	-4.00	-2.00
Empirical	1.85 (2.30)	0.00 (0.00)	-3.76 (4.62)	-1.88 (1.12)
$W^{(1)}-W^{(3)}$	2.53 (3.13)	$4\ 10^{-3}$ (0.08)	-3.87 (5.34)	-1.96 (5.38)
$W^{(2)}-W^{(3)}$	2.11 (2.72)	$-1\ 10^{-3}$ (0.05)	-4.03 (4.19)	-2.02 (4.19)

Table 3 : Empirical and Bootstrap estimators of a , $K'_{1,1}(P)$, $K_{1,1}(P)$, $K'_{1,3}(P)$ for $T(P) = E_P X$, n =100, nboot = 500, mean and standard deviation over 200 replications

Using Theorem 3.5 we obtain confidence intervals for T(F) by Edgeworth expansion inversion. The next tables 4 and 5 show how these intervals perform in terms of coverage probability, respectively for an original level equal to $\alpha = 1\%$ and $\alpha = 5\%$, compared to the asymptotic intervals.

	$P\{ n^{1/2}(\overline{X}-m)/\sigma \leq . \}$		$P\{ n^{1/2}(\overline{X}-m)/S_n \leq . \}$	
n=100	Quant. (α=1%)	Cover. %	Quant. (α=1%)	Cover. %
True	-2.178	1.00	-2.722	1.00
Gaussian	-2.326	0.61	-2.326	2.32
$W^{(1)}-W^{(3)}$	-2.175	1.00	-2.709	0.97
$W^{(2)}-W^{(3)}$	-2.171	1.00	-2.723	1.00

Tables 4 : Coverage probability for the mean and the standardized and the standardized mean, comparison with the gaussian approximation , n = 100 , α = 1%

	$P\{ n^{1/2}(\overline{X}-m)/\sigma \leq . \}$		$P\{ n^{1/2}(\overline{X}-m)/S_n \leq . \}$	
n=100	Quant. (α=5%)	Cover. %	Quant. (α=5%)	Cover. %
True	-1.586	5.0	-1.932	5.0
Gaussian	-1.644	4.3	-1.644	8.4
$W^{(1)}-W^{(3)}$	-1.573	5.1	-1.848	6.0
$W^{(2)}-W^{(3)}$	-1.582	5.0	-1.859	5.8

Table 5 : Coverage probability based on the bootstrap estimates for the mean and the standardized mean, comparison with the gaussian approximation , $n = 100$, $\alpha = 5\%$

In both cases it is clear that the asymptotic performs very badly, particularly when we standardize the empirical mean by a variance estimator. The relative error on the level is close to 100% . The inversion with the Bootstrap yields very accurate confidence intervals for $n^{1/2}(\overline{X}-m)/\sigma$ and a good correction in the studentized case. It is clear that an Edgeworth inversion with the empirical estimators would have yielded similar results. This gives however some indications on how the method may perform in more complex situations.

b) Simulation 2 : $T(F) = \int F_1 dF_2$.

This section presents simulations for the functional $\int F_1 dF_2$. We choose $F_1 = \mathcal{U}[\bar{a},\bar{b}]$ and $F_2 = \mathcal{U}[\bar{c},\bar{d}]$ with $\bar{a} < \bar{b} < \bar{c} < \bar{d}$. Consider two independent i.i.d. samples $X_1, \ldots, X_n$ and $Y_1, \ldots, Y_n$ with distributions F_1 and F_2 . The sample size is equal to $n = 50$, the number of Bootstrap resamplings is nboot = 500. We make nrep = 100 replications of the procedure. Since the two samples are independent, the statistic $T(F_n)$ associated to the functional is no more than the Wilcoxon statistic.

One may easily check that in this setting we have

$$T(F) = ((\bar{b}-\bar{a})^2 - (\bar{c}-\bar{a})^2)/(2\,(\bar{d}-\bar{c})(\bar{b}-\bar{a})) + (\bar{d}-\bar{b})/(\bar{d}-\bar{c}) \quad \text{and}$$

$$s^2(F) = S_1 - T(F)^2 + S_2 - \tilde{T}(F)^2 ,$$

where

$$\tilde{T}(F) := \int F_2 dF_1 = 1 - T(F) ,$$

$$S_1 := \{(\bar{b}-\bar{a})^3 - (\bar{c}-\bar{a})^3\}/(3\,(\bar{b}-\bar{a})^2(\bar{d}-\bar{c})) + (\bar{d}-\bar{b})/(\bar{d}-\bar{c}) ,$$

$$S_2 := (\bar{b}-\bar{c})^3/(3(\bar{d}-\bar{c})^2(\bar{b}-\bar{a})) .$$

Moreover, we have

$$K_{1,1}(F) = 0 , \quad \text{(because } T(F_n) \text{ is an unbiased U-statistic)}$$

$$E_F T^{(1)}(X,F)^3 = ET_1^3 - 3\,T(F)\,S_1 + 2T(F)^3 + ET_2^3 - 3\tilde{T}(F)\,S_2 + 2\,\tilde{T}(F)^3$$

with

$$ET_1^3 = ((\bar{b}-\bar{a})^4 - (\bar{c}-\bar{a})^4)/\{\,4\,(\bar{b}-\bar{a})^3(\bar{d}-\bar{c})\} + (\bar{d}-\bar{b})/(\bar{d}-\bar{c}) ,$$

$$ET_2^3 = (\bar{b}-\bar{c})^4/\{4(\bar{d}-\bar{c})^3(\bar{b}-\bar{a})\} .$$

We easily calculate

$$E_F T^{(1)}(X,F)T^{(1)}(Y,F)T^{(2)}(X,Y,F) = -2\,ET + 2\,\tilde{T}(F)\,T(F)^2$$

$$+ 2\,T(F)(\tilde{T}(F) - S_2) - 2\,\tilde{T}(F)T(F)$$

with

$$ET = \Big((\bar{b}-\bar{a})^2(\bar{b}-\bar{c})^2/2 - ((\bar{b}-\bar{a})^4-(\bar{c}-\bar{a})^4)/4 + (\bar{c}-\bar{a})((\bar{b}-\bar{a})^3-(\bar{c}-\bar{a})^3)/3$$

$$+ (\bar{b}-\bar{c})^2(\bar{d}-\bar{b})(\bar{b}-\bar{a})\Big)/(2(\bar{d}-\bar{c})^2(\bar{b}-\bar{a})^2) .$$

In the following we choose : $\bar{a} = 0$, $\bar{b} = 2$, $\bar{c} = 1$, $\bar{d} = 4$, yielding

$$T(F) = 0.9166 ,$$

$$s^2(F) = 0.032407 ,$$

$$a = E_F T^{(1)}(X,F)^3 = -3.247\ 10^{-3} ,$$

$$ETTT := E_F T^{(1)}(X,F)T^{(1)}(Y,F)T^{(2)}(X,Y,F) = -8.4876\ 10^{-3} ,$$

$K_{1,3}(F) = -4.9597$.

Other simulations results with different paramaters may be obtained on request.

	$s^2(P)$	$K_{1,3}(P)$	a	$K_{1,1}(P)$	$K'_{1,3}(P)$	$K'_{1,1}(P)$
True	0.0324	-4.959	-3.4e-3	0e-0	3.505	5.556
Empiric.	0.03267	-3.650	-3.3e-3	8e-3	2.594	4.230
$W^{(1)}-W^{(3)}$	0.0324*	-3.208*	-4.1e-3	6e-3	2.568	4.182
$W^{(2)}-W^{(3)}$	0.0325**	-3.198**	-2.8e-3	14e-3	2.400	3.816

Table 6 : Empirical and Bootstrap estimates of the variance and the terms in The Edgeworth expansion, n = 100. (*) is obtained only with $W^{(1)}$ (**) obtained with $W^{(2)}$

Table 6 gives some indications on the performance of the boostrap estimators studied in this first chapter. None of the estimators (even the empirical one) are really satisfactory. They just give the right order. However, it is sufficient to obtain rather accurate confidence intervals as can be seen in table 7 .

Figure 2 compares the left tail of the different Bootstrap distributions with the true and asymptotic ones. One more time, the weighted Bootstrap with weights $W^{(2)}$ gives a more accurate approximation of the true distribution.

The conclusions for the functional $\int F_1 \, dF_2$ are similar to those in the case of the mean. The explicit Edgeworth expansion inversion with the generalized Bootstrap estimators yields accurate confidence intervals. We shall see in next part that the choice of the kurtosis of the weight can yield an automatic third order accuracy of the weighted Bootstrap and give better results. Other simulations in these cases will be given in part II.

	$H_n(.,P)$		$K_n(.,P)$	
n=100	Quant. ($\alpha=5\%$)	Cover. %	Quant. ($\alpha=5\%$)	Cover. %
True	-1.80	5.0	-1.42	5.0
Gaussian	-1.64	6.4	-1.64	9.8
$W^{(1)}-W^{(3)}$	-1.74	5.3	-1.40	4.9
$W^{(2)}-W^{(3)}$	-1.74	5.3	-1.42	5.1

Table 7 : Coverage probability based on the bootstrap estimates of the cumulants. Comparison with the true and the asymptotic intervals

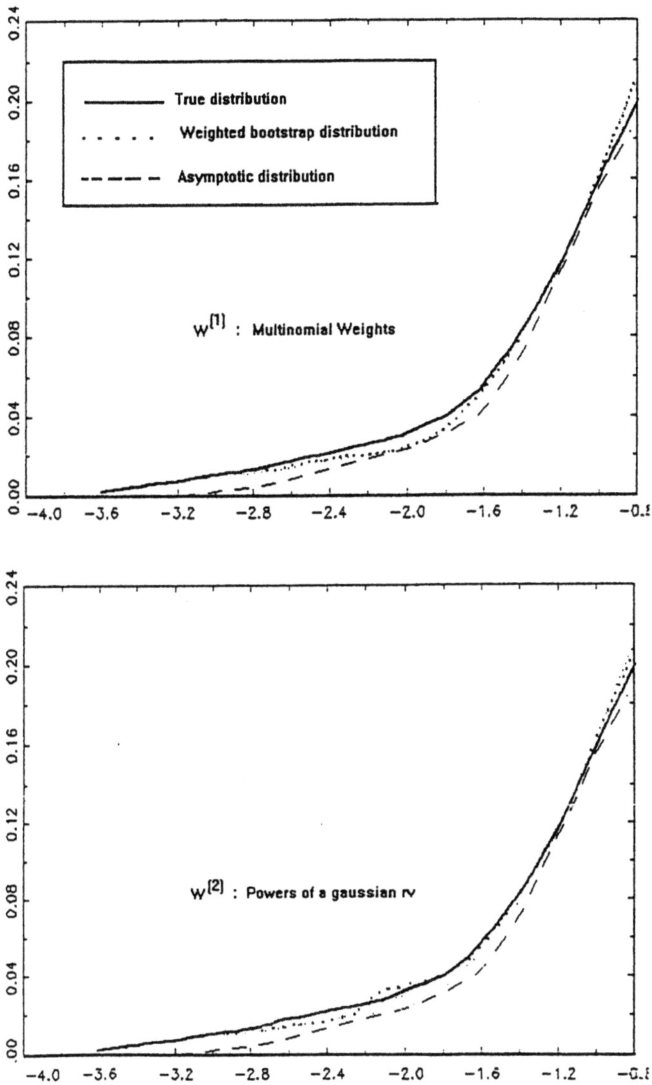

Figure 2 : Comparison of the Bootstraps distributions with the true and the asymptotic ones in the case $\int F_1 dF_2$.

CHAPTER II

HOW TO CHOOSE THE WEIGHTS

1) Introduction

The aim of this chapter is to discuss the practical choice of the random weights $W_{i,n}$ which are used in the generalized bootstrap. Instead of building a complete mathematical theory, we will give some theoretical indications and illustrate them by some simulations.

In the first chapter, the only constraints that we had on the weights were basically on the value of the first moment (see I.1.1) and the existence of a second moment. However, it would be extremely surprising that all the weights with $E\,W_{i,n} = 1/n$ lead to the same results in practice. It turns out that various differences appear, and that generally, the choice of the weights must be data driven and also dependent on the problem. We shall not use the same weights to bootstrap a "smooth" functional $T(P_n)$ or to bootstrap the extreme values, or to bootstrap the mean in the nongaussian stable limiting distribution case.

The organization of this chapter is the following. In section 2 we deal with almost sure results for the weighted bootstrap. We prove that the results of Chapter I holds almost surely along the sequence $(X_i)_{i\geq 1}$ when the weights are generated by i.i.d. r.v.'s (consequence of Proposition 2.1). This allows us to drop assumption (I.1.3) and choose the distribution of the weights as a function of the data. Section 3 provides indications on the choice of i.i.d. weights for the mean. We quote the results of Haeusler, Mason and Newton (1992) who show that with properly chosen weights, the generalized bootstrap is as good as the ordinary one (Theorem 3.1.1). Haeusler (1992) obtains better confidence intervals by choosing appropriate weights (see section 3.2). These results are extended in sections 4 and 5 to some regular functionals $T(P)$. Section 4 is devoted to the Edgeworth expansion for the generalized bootstrapped studentized statistic up to a $o(n^{-1})$! (Theorem 4.1.) This gives conditions on the weights to obtain second order correct confidence intervals (Corollary 4.1). The analogous results to those of Haeusler are obtained in section 5 for differentiable statistics. The main difficulty is to handle the nonlinearities of the functional $T(.)$. The theoretical results are illustrated by some simulations in subsection 5.2. In section 6, we look for large deviations of the generalized bootstrapped p.m.. Theorem 6.3 shows that in the large deviation sense, the classical bootstrap cannot really be improved by the generalized one!

In all this chapter, we shall assume that the functional is real valued. When we study the mean, this means that the sample $X_1, \ldots, X_n$ is real valued. We also investigate the generalized bootstrap for the following special type of weights : let $(\mathcal{Y}_n)_{n\geq 1}$ be a sequence of distribution functions on $\mathbb{R}$, which may depend on the sample $X_1, \ldots, X_n$. Then, define a triangular array

(1.1) $$\{ Y_{i,n} : 1 \leq i \leq n , n \geq 1 \}$$

of r.v.'s such that each row is a vector of i.i.d. r.v.'s with distribution $\mathcal{Y}_n$. Further, let

(1.2) $$n\overline{Y}_n := Y_{1,n} + \ldots + Y_{n,n}$$

and define the weights (called iid-generated weights) (see Lo (1991))

(1.3) $$W_{i,n} = Y_{i,n} / n\overline{Y}_n .$$

We can define the corresponding bootstrapped empirical measure $P_{W,n}$.

When using this scheme we can drop assumption (I.1.3) since $\mathcal{Y}_n$ may be data dependent. It turns out that we can also strengthen our previous results in proving some a.s. statements. For practical purposes, the fact that $\mathcal{Y}_n$ may be data dependent is of real importance. We shall see in the next sections that to obtain nice applications of this generalized bootstrap, we must choose the weights data dependent.

2) Weights generated from an i.i.d. sequence : almost sure results

To obtain almost sure versions of the theorems of chapter I with the weights defined in (1.3), all what we have to do, is to show that Proposition I.2.2 and particularly (I.2.20) remains valid a.s. along the sample.

We shall assume condition (I.3.5), but it is more convenient to state this in terms of the sequence $Y_{i,n}$. Then, for some suitable r we assume

(2.1) $$Y_{i,n} \geq 0 ,$$

(2.2) $$\limsup_{n\to\infty} E|Y_{i,n}|^r < \infty , \text{ and}$$

(2.3) $$\limsup_{n\to\infty} E\, \bar{Y}_n^{-r} < \infty .$$

Now, we can state the analogue of Proposition I.2.2 .

<u>Theorem 2.1</u> : **Let** $\rho > 2$ **and assume that**

. (I.2.10), (I.2.13), (I.2.14) **and** (2.1) **hold,**

. (2.2) **holds for** $r = D \vee 2 \vee 3\rho$,

. (2.3) **holds for** $r = 3\rho$.

Then, the following bound holds for any $c > 0$ **and any** n **larger than some deterministic** N_1 **(depending on** c **) :**

$$E_W(d_{\mathcal{H}}(P_{W,n},P_n)^{\rho} \mid P_n)$$

$$\leq K_1\, n^{-\rho/2}(n^{-1}\Sigma_{i=1}^n H(X_i)^2)^{\rho/2}$$

$$+ K_2\, (n^{-1}\log n)^{\rho/2}(n^{-1}\Sigma_{i=1}^n H(X_i)^{2\rho/(\rho-1)})^{(\rho-1)/2}$$

$$+ K_3\, (\log n)^{(\rho-2)/2}\, n^{-c}\, n^{-\rho/2}\, (n^{-1}\Sigma_{i=1}^n H(X_i)^4)^{\rho/4}$$

$$+ K_4\, n^{-\rho/2}\, (n^{-1}\Sigma_{i=1}^n H(X_i)^2)^{1/2}$$

$$+ K_5\, (\log n)^{(\rho-2-D)/2}\, n^{-c}\, n^{(D-\rho)/2}\, (n^{-1}\, \Sigma_{i=1}^n H(X_i)^2)^{D/2}$$

$$(n^{-1}\, \Sigma_{i=1}^n H(X_i)^4)^{(\rho-D)/2}$$

where $K_1, \ldots, K_5$ **are constants depending only on the conditions** (2.2) **and** (2.3).

Therefore, under the additional condition

(2.4) $$E\, H(X_i)^{\rho\vee 2} < \infty ,$$

we have the order

$$(2.5)\qquad E_W\big(\, d_{\mathcal{H}}(P_{W,n},P_n)^{\rho} \mid P_n\,\big) = O(n^{-\rho/2}(\log n)^{\rho/2}) \quad \text{a.s. and}$$

$$(2.6)\qquad d_{\mathcal{H}}(P_{W,n},P_n) = O_W(n^{-1/2}(\log n)^{1/2}) \quad \text{a.s.}$$

Proposition 2.1 : **Under the same conditions as in Theorem 2.1 and (2.4) with** $\rho > 2(D+2)$ **we have**

$$(2.7)\qquad d_{\mathcal{H}}(P_n,P) = O(n^{-1/2}(\log n)^{1/2}) \quad \text{a.s.}$$

The results of Theorem 2.1 and (2.5)-(2.7) lead immediately to the almost sure version of the theorems of the first chapter.

Under the assumptions of Theorem 2.1 with $\rho = 2$ (and since c may be taken as large as we want), under $E\,H(X_i)^2 < \infty$, it is easy to see that Theorem I.3.1 holds, with convergence L^* replaced by L^{as} , convergence in distribution conditionally on the sample, a.s. along the sequence.

Under the additional assumptions of Theorem 2.1 with $\rho = 6$, Theorems I.3.2, I.3.3 and I.3.4 hold a.s. along the sample.

3) Best weights for the bootstrap of the mean via Edgeworth expansion

In this paragraph, we shall focus on some properties of the weighted bootstrap, when the weights are chosen i.i.d. This will give us some information on how the weights should be designed.

3.1) Second order correction

To restate our basic question for the mean, we first recall the notations, $\overline{X}_n := n^{-1}\,\Sigma_{i=1}^n X_i$ for the usual empirical mean, and $\overline{X}_{Y,n} := n^{-1}\,\Sigma_{i=1}^n Y_{i,n}X_i$ for the bootstrapped mean with i.i.d. weights $(Y_{1,n}, \ldots, Y_{n,n})$. Now, the question is : what are the differences between all the $\overline{X}_{Y,n}$ when the distribution of the weights varies, and which one is the best?

The first idea is of course to look how accurate is the bootstrapped approximation. We assume that the X_i's admit a variance and define

$$\sigma^2 := \text{Var } X_i < \infty .$$

Let us denote

$$G_n(x) := P\{ n^{1/2}(\overline{X}_n - EX_1)/\sigma \le x \}$$

the true and unknown d.f. of the mean. Define the bootstrapped d.f.

$$G_{Y,n}(x) := P\{ n^{1/2}(\overline{X}_{Y,n} - \overline{X}_n\overline{Y}_n)/\hat{\sigma}_n\sigma_{Y,n} \le x \mid P_n \} \tag{3.1}$$

where P_n denotes the empirical measure and $\hat{\sigma}_n^2$ the empirical variance,

$$\hat{\sigma}_n^2 := n^{-1} \Sigma_{i=1}^n (X_i - \overline{X}_n)^2 ,$$

and $\sigma_{Y,n}$ is the standard deviation of $Y_{i,n}$, i.e. $\sigma_{Y,n}^2 = \text{Var}(Y_{i,n} | P_n)$.

Representation (3.1) allows $\overline{Y}_n$ to be different from 1, but it is natural to assume that

$$E\, Y_{i,n} = 1 .$$

See remark I.3.2 for a discussion on such a normalization. The skewness and kurtosis of $Y_{i,n}$ are respectively denoted by

$$\beta_{Y,n} := \text{Skew}(Y_{i,n}) = E(Y_{i,n} - 1)^3/\sigma_{Y,n}^3 \quad \text{and}$$

$$K_{Y,n} := \text{Kur}(Y_{i,n}) = E(Y_{i,n} - 1)^4/\sigma_{Y,n}^4 .$$

Respectively the empirical skewness and empirical kurtosis are given by

$$\hat{\beta}_{X,n} := n^{-1} \Sigma_{i=1}^n (X_i - \overline{X}_n)^3 / \hat{\sigma}_n^3$$

$$\hat{K}_{X,n} := n^{-1} \Sigma_{i=1}^n (X_i - \overline{X}_n)^4 / \hat{\sigma}_n^4$$

The Edgeworth expansion for $G_{Y,n}$ up to a $o(n^{-1})$ term is given by Haeusler, Mason and Newton (1992) who prove the following.

Theorem 3.1 : **Assume that the r.v.'s $Y_{i,n}$ are i.i.d. $\mathcal{Y}_n$ and that the sequence $\{Y_{i,n}^4 : n\geq 1\}$ is uniformly integrable under $\mathcal{Y}_n$ and**

$$\limsup_{|t|\to\infty} \sup_{n\geq 1} |E(\exp(itY_{1,n}) \mid P_n)| < 1 .$$

Then, with probability one,

$$\sup_{t\in\mathbb{R}} | G_{Y,n}(t) - \mathcal{E}^Y_{2,G}(t) | = o(n^{-1})$$

as $n\to\infty$, where

$$\sigma_{Y,n}^2 := \mathrm{Var}(Y_{1,n} \mid P_n) , \quad \text{and}$$

$$\mathcal{E}^Y_{2,G}(t) := \Phi(t) - n^{-1/2}(1/6)(t^2-1)\phi(t)\hat{\beta}_{X,n}\beta_{Y,n}$$

$$- (1/24n)\ (t^3-3t)\phi(t)\hat{K}_{X,n}(K_{Y,n} - 3)$$

$$- (1/72n)\ (t^5 - 10t^3 + 15t)\phi(t)\hat{\beta}_{X,n}^2\ \beta_{Y,n}^2 ,$$

where $\hat{\beta}_{X,n}$, and $\hat{K}_{X,n}$ are respectively the skewness and kurtosis of X_i under P_n ; $\beta_{Y,n}$ and $K_{Y,n}$ are those of $Y_{i,n}$ conditioned on P_n .

Since under suitable conditions on the X_i's , $\hat{\beta}_{X,n}$ and $\hat{K}_{X,n}$ converge to their theoretical values

$$\beta_X := \mathrm{Skew}(X_i) \quad \text{and} \quad K_X := \mathrm{Kur}(X_i)$$

we see that the bootstrapped distribution can fit G_n only up to a $O_p(n^{-1})$ term by choosing the distribution $\mathcal{Y}_n$ appropriately (see Haeusler, Mason and Newton (1992) for more details).

The main point is that, if we want to fit G_n appropriately, for instance to construct a one sided confidence interval for EX_1 , when σ is known, we have to choose the first three moments of the weights $\{ Y_{i,n} : 1 \leq i \leq n \}$. More precisely, if we choose $\beta_{Y,n} = 1 + o(1)$ then under the assumptions of Theorem 3.1, we have that :

$$\sup_{t\in\mathbb{R}} |P\{n^{1/2}(\overline{X}_{Y,n}- \overline{X}_n)/(\sigma_{Y,n}\hat{\sigma}_n) \leq x\} - P\{n^{1/2}(\overline{X}_n- EX_1)/\sigma \leq x\}|$$

$$= o_p(n^{-1/2}) \quad \text{as} \quad n \to \infty \ .$$

On the other hand, if we want to construct two-sided confidence intervals, we have to choose the weights such that we can match the Edgeworth expansion of

$$H_n^T(x) := P\{ n^{1/2} |\overline{X}_n - EX_1| / \sigma \leq x \} = B_{2,n}(x) + o(n^{-1})$$

and

$$H_{Y,n}^T(x) := P\{ n^{1/2} |\overline{X}_{Y,n} - \overline{X}_n| / (\sigma_{Y,n} \hat{\sigma}_n) \leq x \} = B_{2,n}^Y(x) + o(n^{-1})$$

with

$$B_{2,n}(x) = 2\Phi(x) - 1 - (12n)^{-1} (K_X - 3) (x^3 - 3x) \phi(x)$$

$$- (36n)^{-1} \beta_X^{\ 2} (x^5 - 10x^3 + 15x) \phi(x)$$

and

$$B_{2,n}^Y(x) = 2\Phi(x) - 1 - (12n)^{-1} \hat{K}_{X,n} (K_{Y,n} - 3) (x^3 - 3x) \phi(x)$$

$$- (36n)^{-1} \hat{\beta}_{X,n}^{\ 2} \beta_{Y,n}^{\ 2} (x^5 - 10x^3 + 15x) \phi(x)$$

Thus, in this case we have to choose the first four moments of the weights. If we choose $\beta_{Y,n} = 1 + o(1)$ and $K_{Y,n} = 4 - 3/\hat{K}_{X,n} + o(1)$ then we have

$$\sup_{x \in \mathbb{R}} |H_n^T(x) - H_{Y,n}^T(x)| = o_P(n^{-1}) \quad \text{as} \quad n \to \infty \ .$$

Since the odd terms in the Edgeworth expansion of $H_n^T(.)$ and $H_{Y,n}^T(.)$ vanish by symmetry of the distribution, the last approximation is valid up to $o_P(n^{-3/2})$ and even $O_P(n^{-2})$, if some additional conditions on the moments of the X_i's and $Y_{i,n}$'s are satisfied.

Similar results may also be obtained for the studentized mean, which allows the construction of more accurate confidence intervals for μ , without assuming any knowledge on σ .

The main point here is that in any case, one must only designate the first four moments of the weights. A practical way to generate i.i.d weights with given first four moments is given in Devroye (1986). However the given choice for $K_{Y,n}$ is made possible only under some constraints on the distribution of the X_i's (see Hauesler, Mason and Newton (1992)), typically if the distribution of the X_i's is unimodal. This appears to be a serious drawback of this i.i.d. weighted sheme.

3.2) Coverage probability

The next step has been realized by Haeusler (1992) who looks at the coverage probability. He assumes that

$$\mathcal{Y}_n \text{ is a continuous distribution.}$$

For a given level α, we define $c_{Y,\alpha,n}$ by

$$G_{Y,n}(c_{Y,\alpha,n}) = \alpha$$

Haeusler (1992) expands the (unconditional) coverage probability

$$(3.2.1) \qquad p_n(\alpha) := P\{ n^{1/2}(\bar{X}_n - EX_1)/\sigma \le c_{Y,\alpha,n} \}$$

and shows that

$$p_n(\alpha) = \alpha + n^{-1}\, u_\alpha \phi(u_\alpha)\, c_{\alpha,n}(h,k) + o(n^{-1})$$

with

$$c_{\alpha,n}(h,k) := \beta_X\, (u_\alpha^2-1)\, ((6u_\alpha)^{-1}h - 4^{-1}\beta_X)$$

$$+ 24^{-1}(K_X-3)\,(3u_\alpha^2-1) + K_X\, k\, 24^{-1}(u_\alpha^2-3) \quad \text{and}$$

$$u_\alpha := \Phi^{-1}(\alpha)$$

when we choose the skewness and the kurtosis of $\{ Y_{i,n} : 1 \le i \le n \}$ such that $\beta_{Y,n} = 1 + n^{-1/2}h\,(1+o(1))$ and $K_{Y,n} = k + 3$, for some $k > 1$.

It follows that we have to find h and k, which may depend on the data and α, such that $c_{\alpha,n}(h,k) = 0$ under the constraint (due to the fact that Hölder's inequality implies $|\beta_{Y,n}|^{1/3} \le |K_{Y,n}|^{1/4}$), $|1 + n^{-1/2}h| \le (k+3)^{3/4}$ or at least to minimize $c_{\alpha,n}(.)$ under the given constraint. The solution of such program essentially depends on the value of β_X, K_X and u_α.

If $\beta_X \ne 0$ and $u_\alpha^2 \ne 1$ then we can always fix the value of $k > 1$ and then solve $c_{\alpha,n}(h,k) = 0$ to find a solution for h as a function of β_X and K_X and then use the empirical version of h say h_n. For n sufficiently large the constraint will be automatically satisfied. However, we shall see in the simulations that for small and fixed values of n, some practical problems arise (see the simulations results in 5.3).

Now if $\beta_X = 0$ or if $u_\alpha^2 = 1$, then we must minimize

$$c_{\alpha,n}(0,k) = 24^{-1} (K_X-3) (3u_\alpha^2-1) + 24^{-1} K_X k (u_\alpha^2-3) ,$$

under $k > 1$. However according to the value of K_X it is not always possible to make this term vanish. For instance, assuming $u_\alpha < -3^{-1/2} \simeq -0.57$ and $K_X \le 3$ it is easy to see that $c_{\alpha,n}(0,k) = 0$ has no solution. Thus in that case the best solution is to choose $k = 1$, $h = 0$ and $p_n(\alpha)$ becomes

$$p_n(\alpha) = \alpha + n^{-1} u_\alpha \phi(u_\alpha) c_{\alpha,n}(0,1) = \alpha + O(n^{-1})$$

A third order correction may be obtained by correcting directly the bootstrap quantile by this quantity (which is the original idea of Abramovitch and Singh (1985)).

This last case (i.e. $\beta_X = 0$ or $u_\alpha^2 = 1$) seems to be a rather extreme case. However, notice that in the gaussian case and the uniform case, we can not choose the weights so as to obtain a third order correction with a weighted bootstrap! In that case, the 'optimal' weighted bootstrap performs worse than the asymptotic approximation.

If we use Efron's bootstrap, the same procedure enables us to define a bootstrapped critical value $c^*_{\alpha,n}$ and Hall (1986) proved that $p^*_n(\alpha) = \alpha + cn^{-1}$ where $c \neq 0$. Hence, Haeusler's result proves that in most cases the generalized bootstrap can do much better than Efron's bootstrap, provided the distribution of the weights is suitably chosen.

4) Choice of the weights for a general functional via Edgeworth expansion

The aim of this paragraph and of the next one is to generalize the results of Haeusler, Mason and Newton (1992), Haeusler (1992) to some general third order differentiable functionals. It is clear that such a task is made more difficult by the presence of non-linearities but may be more useful in practical situations. Moreover, this would be quite easy if the random weights $W_{i,n}$ were i.i.d. as the one we took in sections 2 and 3. However, Remark I.3.2 shows that i.i.d. weights lead to some difficulties, and that it is natural to assume that (I.1.1) is satisfied. A convenient way to achieve (I.1.1) is to choose i.i.d. generated weights as in (1.3). Section I.5 gives a practical way to generate such exchangeable weights with given first four moments.

4.1) Edgeworth expansion up to $o(n^{-1})$ for a third order differentiable functional

Consider the general problem of choosing the weights for a differentiable functional. The preceeding results suggest that the weights should depend on the moments of the gradients. To obtain such results we first begin by recalling the Edgeworth expansion for a third order differentiable functional (see Appendix 8 or Chapter 4 of Bertail (1992)) . The main tool to obtain this result is the delta-method combined with the Edgeworth expansion of a U-statistics of degree 3. The technical conditions under which such an expansion holds may be derived in the same manner as in Theorem I.3.3. However such a task may be rather tedious and is not within the scope of this chapter.

We have, with the same notations as in section I.3,

(4.1) $$H_n(x,P) = P\{ n^{1/2} (T(P_n)-T(P))/S(P) \le x \}$$

$$= \mathcal{E}_{2,H}(x,P) + o(n^{-1}) ,$$

with

(4.2) $$\mathcal{E}_{2,H}(x,P) := \Phi(x) - n^{-1/2}\left(\frac{K_{1,1}(P)}{2} + \frac{K_{1,3}(P)}{6} (x^2-1) \right) \phi(x)$$

$$-n^{-1} \left(\frac{K_{2,2}(P)}{2} x + \frac{K_{2,4}(P)}{24} (x^3-3x) + \frac{K_{2,6}(P)}{72} (x^5-10x^3+15x) \right) \phi(x)$$

and with the coefficients

(4.3) $$S(P)\, K_{1,1}(P) := E_P\, T^{(2)}(X_1,X_1,P)$$

$$S(P)^3\, K_{1,3}(P) := E_P\, T^{(1)}(X_1,P)^3$$

$$+ 3\, E_P\, T^{(2)}(X_1,X_2,P)T^{(1)}(X_1,P)T^{(1)}(X_2,P)$$

$$S(P)^2\, K_{2,2}(P) := 4^{-1}\,(E_P\, T^{(2)}(X_1,X_1,P))^2 + 2^{-1}\, E_P\, T^{(2)}(X_1,X_2,P)^2$$

$$+\, E_P\, T^{(1)}(X_1,P)T^{(2)}(X_1,X_1,P) + E_P\, T^{(1)}(X_1,P)T^{(3)}(X_1,X_2,X_2,P)$$

$$S(P)^4\, K_{2,4}(P) := 2\, S(P)^4 K_{1,1}(P)\, K_{1,3}(P) + E_P T^{(1)}(X_1,P)^4 - 3S(P)^4$$

$$+\, 12\, E_P\, T^{(1)}(X_1,P)\, T^{(1)}(X_2,P)^2\, T^{(2)}(X_1,X_2,P)$$

$$+\, 12\, E_P\, T^{(1)}(X_2,P)\, T^{(1)}(X_3,P)\, T^{(2)}(X_1,X_2,P)\, T^{(2)}(X_1,X_3,P)$$

$$+\, 4\, E_P\, T^{(1)}(X_1,P)\, T^{(1)}(X_2,P)\, T^{(1)}(X_3,P)\, T^{(3)}(X_1,X_2,X_3,P)$$

$$K_{2,6}(P) := K_{1,3}(P)^2\, .$$

(see also Withers (1983) for some formal calculations).

4.2) Edgeworth expansion up to $o(n^{-1})$ for the weighted version

The next step is to obtain an Edgeworth expansion for

$$D_{W,n} := n^{1/2}(T(P_{W,n})-T(P_n))\, /\, S(P) \quad \text{up to} \quad o(n^{-1})\, .$$

Clearly, we need to assume that $E\, Y_{i,n} \neq 0$, and there is no loss of generality to assume

$$E\, Y_{i,n} = 1\, .$$

To calculate the Edgeworth expansion for $D_{W,n}$ up to $o(n^{-1})$, we need to apply Taylor's formula on $T(.)$ to expand $D_{W,n}$ up to $o(n^{-1})$. In this expansion, we have some terms depending on the first three gradients $T^{(1)}(.,P)$, $T^{(2)}(.,P)$ and $T^{(3)}(.,P)$, which are assumed to exist. We also need to control the remainder term in the Taylor expansion (which we have done in chapter I in a simpler case). Since the purpose of these calculations is to give some indication on how one should choose the weights, we do not give an impressive list of conditions on T which would ensure the validity of the Edgeworth expansion. We shall just say "under suitable regularity conditions" to mean that :

(4.4) the functional is third order differentiable and may be approximated by its Taylor expansion up to $o(n^{-1})$ in all the calculations

(combine chapter I and the proofs of the results of this section to get some conditions...)

(4.5) The variance estimator S_n^2 defined in (I.3.3) may be approximated by its expansion up to $o(n^{-1})$ in all the proofs.

These regularity conditions imply that the moments of the weights exist up to a certain order. The exact (optimal ?) assumptions are extremely difficult to obtain, but they are satisfied for instance if $\mathcal{Y}_n$ has uniformly (in n) compact support or uniformly exponential tail.

We need the following notations :

(4.6)
$$\tau_i := T^{(1)}(X_i,P) - n^{-1} \Sigma_{j=1}^n T^{(1)}(X_j,P)$$
$$+ n^{-1} \Sigma_{j=1}^n T^{(2)}(X_i,X_j,P) - n^{-2} \Sigma_{k=1}^n \Sigma_{j=1}^n T^{(2)}(X_k,X_j,P)$$
$$+ 2^{-1} n^{-2} \Sigma_{k=1}^n \Sigma_{j=1}^n T^{(3)}(X_i,X_j,X_k,P)$$

$$\tau_{i,j} := T^{(2)}(X_i,X_j,P)$$
$$- 2^{-1} \left(n^{-1} \Sigma_{k=1}^n T^{(2)}(X_i,X_k,P) + n^{-1} \Sigma_{k=1}^n T^{(2)}(X_j,X_k,P) \right)$$
$$+ n^{-1} n^{-1} \Sigma_{k=1}^n T^{(3)}(X_i,X_j,X_k,P)$$
$$- \left(T^{(1)}(X_i,P) + T^{(1)}(X_j,P) - 2\, n^{-1} \Sigma_{k=1}^n T^{(1)}(X_k,P) \right),$$

$$\tau_{i,j,k} := T^{(3)}(X_i,X_j,X_k,P)$$
$$- 2 \left(T^{(2)}(X_i,X_j,P) + T^{(2)}(X_i,X_k,P) + T^{(2)}(X_j,X_k,P) \right)$$
$$+ 2 \left(T^{(1)}(X_i,P) + T^{(1)}(X_j,P) + T^{(1)}(X_k,P) \right),$$

$$\sigma_{\tau,n}^2 := n^{-1} \Sigma_{i=1}^n \tau_i^2 .$$

Then we define

$$\overline{\tau_1^p} := n^{-1} \Sigma_{i=1}^n \tau_i^p / \sigma_{\tau,n}^p ,$$

$$\overline{\tau_{1,2}^p} := n^{-2} \Sigma_{i=1}^n \Sigma_{j=1}^n \tau_{i,j}^p / \sigma_{\tau,n}^p ,$$

$$\overline{\tau_1\tau_2\tau_{1,2}} := n^{-2} \Sigma_{i=1}^n \Sigma_{j=1}^n \tau_i\tau_j\tau_{i,j} / \sigma_{\tau,n}^3 ,$$

and similarly $\overline{\tau_1\tau_2\tau_{1,2}\tau_{2,3}}$ and so on... (for simplicity of notations we have removed the subscript n in the notation $\overline{\tau_1^p}$, etc.).

Then we have the following result.

Proposition 4.1 : **Under the suitable regularity conditions** (4.4)-(4.5) **and if**

(4.7) $$\limsup_{n\to\infty} \limsup_{t\to\infty} |E \exp(itY_{1,n})| < 1$$

we have the following Edgeworth expansion :

(4.8) $$H_{W,n}(x,P) := P\{ n^{1/2}(T(P_{W,n}) - T(P_n))/(\sigma_{\tau,n}\sigma_{Y,n}) \le x \mid P_n \}$$

$$= \mathcal{E}_{2,H}^W(x,P) + o_p(n^{-1})$$

with

$$\mathcal{E}_{2,H}^W(x,P) := \Phi(x) - n^{-1/2}\left(2^{-1}K_{1,1,n}^Y(P) + 6^{-1}K_{1,3,n}^Y(P)(x^2-1)\right) \phi(x)$$

$$-n^{-1} \left(2^{-1} K_{2,2,n}^Y(P)\, x + 24^{-1} K_{2,4,n}^Y(P)\, (x^3-3x)\right.$$

$$\left.+ 72^{-1} K_{2,6,n}^Y(P)\, (x^5 - 10x^3 +15x) \right) \phi(x) + o(n^{-1}) ,$$

and

(4.9) $$K_{1,1,n}^Y(P) := \sigma_{Y,n} \overline{\tau_{1,1}} ,$$

$$K_{1,3,n}^Y(P) := \beta_{Y,n} \overline{\tau_1^3} + 3\, \sigma_{Y,n} \overline{\tau_1\tau_2\tau_{1,2}} ,$$

$$K_{2,2,n}^Y(P) := 4^{-1} \sigma_{Y,n}^2 \overline{\tau_{1,1}}^2 + 2^{-1} \sigma_{Y,n}^2 \overline{\tau_{1,2}^2} + \sigma_{Y,n} \beta_{Y,n} \overline{\tau_1\tau_{1,1}}$$

$$+ \sigma_{Y,n}^2 \overline{\tau_1\tau_{1,2,2}} ,$$

$$K^Y_{2,4,n}(P) := 2\, K^Y_{1,1,n}(P)\, K^Y_{1,3,n}(P) + K_{Y,n}\, \overline{\tau_1^4} - 3$$

$$+ 12\, \beta_{Y,n}\, \sigma_Y\, \overline{\tau_1 \tau_2 \tau^2_{1,2}} + 12\, \sigma^2_{Y,n}\, \overline{\tau_1 \tau_2 \tau_{1,2} \tau_{2,3}}$$

$$+ 4\, \sigma^2_{Y,n}\, \overline{\tau_1 \tau_2 \tau_3 \tau_{1,2,3}}$$

$$K^Y_{2,6,n}(P) := K^Y_{1,3,n}(P)^2 .$$

Observe that the $K^Y_{i,j,n}(P)$'s depend on P through the τ_i , $\tau_{i,j}$ and $\tau_{i,j,k}$. We can infer the second order validity of the iid-generated weighted bootstrap from our Proposition 4.1 and Theorem I.3.3. Indeed, notice first that we have the following results:

(4.10) $$n^{1/2} (\sigma^2_{\tau,n} - s^2(P)) \xrightarrow[n\to\infty]{d} N(0,I(P)) \quad \text{as } n \to \infty ,$$

where $I(P)$ is given in (I.3.10). Then, applying the law of large numbers for U-statistics, we also have

$$\lim_{n\to\infty} K^Y_{1,1,n}(P) = K_{1,1}(P) \quad \text{and}$$

$$\lim_{n\to\infty} K^Y_{1,3,n}(P) = K_{1,3}(P) \quad \text{in probability,}$$

provided that

(4.11) $$\sigma_{Y,n} = 1 + o_P(1) \quad \text{and} \quad \beta_{Y,n} = 1 + o_P(1) .$$

Therefore, we infer the next corollary which proves the second order validity of the weighted bootstrap distribution, yielding second order correct confidence intervals :

Corollary 4.1 : **Assume that the hypotheses of Proposition 3.1 hold as well as (4.11). Then, for any compact set** $C \subset \mathbb{R}$,

(4.12) $$\sup_{x\in C} |H_n(x,P) - H_{W,n}(x,P)| = o_p(n^{-1/2})$$

Under some moments conditions on the first and second gradients, we typically have by the central limit theorem for U-statistics, that

$$K^{Y}_{1,1,n}(P) = K_{1,1}(P) + O_P(n^{-1/2}) ,$$

$$K^{Y}_{1,3,n}(P) = K_{1,3}(P) + O_P(n^{-1/2}) ,$$

so that if

$$\sigma_{Y,n} = 1 + O(n^{-1/2}) \quad \text{and} \quad \beta_{Y,n} = 1 + O(n^{-1/2}) ,$$

then $o_p(n^{-1/2})$ in (4.12) may be replaced by $O_P(n^{-1})$, which is the best rate that we may obtain in that case. Similar results have been obtained for the usual bootstrap in Bertail (1992) for differentiable functionals and Helmers (1991) for U-statistics.

Notice that the result (4.12) does not change if we use an estimate $\tilde{S}^2_n$ of $S^2(P)$ instead of $\sigma_{\tau,n}^{2}$ (which depends on P through $T^{(1)}(.,P)$ and so on) such that

$$(4.13) \qquad \tilde{S}^2_n - \sigma_{\tau,n}^{2} = o_p(n^{-1/2}) \quad \text{as} \quad n \to \infty .$$

For instance, the estimator S^2_n defined in (I.3.3) does satisfy (4.13). However, we shall see further that replacing $\sigma_{\tau,n}^{2}$ by S^2_n induces a small change in the Edgeworth expansion up to $o(n^{-1})$ (see, in the next section, equality 5.6).

Since we may be interested in constructing confidence intervals, it seems more appropriate to deal with the coverage probability as done by Hall (1986) and Haeusler (1992).

5) Coverage probability for the weighted bootstrap of general functionals

5.1) Derivation of the coverage probability

Let $\gamma_{W,n}(\alpha)$ be the α-quantile of the bootstrapped distribution of the statistics $T(P_n)$, defined by

$$(5.1) \qquad P\{ n^{1/2}(T(P_{W,n}) - T(P_n))/(S_n\sigma_{W,n}) \le \gamma_{W,n}(\alpha) \mid P_n \} = \alpha .$$

Remark 5.1 : We assume that the distribution of the weights $W_{i,n}$ is continuous so that $\gamma_{W,n}(\alpha)$ is uniquely defined by (5.1).

Following Hall (1986) and Haeusler (1992), we are interested in calculating the coverage probability

$$(5.2) \qquad P\{ n^{1/2}(T(P_n) - T(P))/S(P) \le \gamma_{W,n}(\alpha) \}$$

and choosing the weights so that this probability is as close as possible to α .

However, (5.2) is not very useful for practical purposes since S(P) is generally unknown. Thus, we shall also consider the coverage probability

$$(5.3) \qquad P\{ n^{1/2}(T(P_n) - T(P))/ S_n \le \gamma_{W,n}(\alpha) \}$$

where S_n is the bootstrapped estimator introduced in (I.3.3).

Throughout this section we keep the same type of iid-generated weights as introduced in the previous section.

Remark 5.2 : One can imagine to use different weights in (I.3.3) to calculate S_n and in (5.1) to calculate the α-quantile $\gamma_{W,n}$. This would imply a form of double bootstrap. It is implicit in the formulation of (5.1) that the weights we use to calculate S_n and generate the bootstrapped distribution in (5.1) have the same distribution and are eventually based on the same r.v.'s. Computationally, this last proposal is extremely time saving. We just have to generate the distribution of $D_{W,n}$ conditionally on P_n and standardize it to an unit variance.

To make a first step towards the approximation of the coverage probabilities (5.2) and (5.3) we shall assume

$$E(nW_{i,n}-1)^2(nW_{2,n}-1)^2 = \sigma_{W,n}^4(1+o(1)) \quad \text{as} \quad n\to\infty \quad \text{and}$$

$$\sigma_{W,n}^2 = \sigma_{Y,n}^2 - n^{-1}(3\sigma_{Y,n}^4 - \sigma_{Y,n}^2 - 2\sigma_{Y,n}^3\beta_{Y,n})\ (1+o(1)) \quad \text{as} \quad n\to\infty$$

(see Appendix I for these assumptions). Using the Edgeworth expansion (4.8), we obtain the following Edgeworth expansion :

$$P\{\, n^{1/2}(T(P_{W,n}) - T(P_n))/\,(S_n\sigma_{W,n}) \le x \mid P_n \,\}$$

(5.6)
$$= \Phi(x) - n^{-1/2}\left(\frac{K^Y_{1,1,n}(P)}{2} + \frac{K^Y_{1,3,n}(P)}{6}\,(x^2-1) \right) \phi(x)$$

$$- n^{-1}\left(\frac{K^Y_{2,4,n}(P)}{24}\,(x^3-3x) + \frac{K^Y_{2,6,n}(P)}{72}\,(x^5-10x^3+15x) \right) \phi(x)$$

$$+ o(n^{-1}) \; .$$

Similarly to Hall (1983, 1986) and Haeusler (1992) we can obtain an approximation for $\gamma_{n,W}(\alpha)$ by inverting the Edgeworth expansion (5.6). Denoting

(5.7)
$$c_{W,n}(\alpha) := u_\alpha + n^{-1/2} \left(\frac{K^Y_{1,1,n}(P)}{2} + \frac{K^Y_{1,3,n}(P)}{6}\,(u_\alpha^{\,2}-1) \right)$$

$$+ n^{-1}\left(\frac{K^Y_{2,4,n}(P)}{24}\,(u_\alpha^{\,3}-3u_\alpha) + \frac{K^Y_{2,6,n}(P)}{72}\,(u_\alpha^{\,5}-10u_\alpha^{\,3}+15u_\alpha) \right)$$

$$+ n^{-1} \left(\frac{K^Y_{1,3,n}(P)}{3} - \frac{K^Y_{1,1,n}(P)}{4} - \frac{K^Y_{1,3,n}(P)}{12}(u_\alpha^{\,2}-1) \right)$$

$$\left(\frac{K^Y_{1,1,n}(P)}{2} + \frac{K^Y_{1,3,n}(P)}{6}\,(u_\alpha^{\,2}-1) \right) u_\alpha \; ,$$

we have the Cornish-Fisher expansion

(5.8)
$$\gamma_{W,n}(\alpha) = c_{W,n}(\alpha) + o_p(n^{-1}) \; .$$

The next step is to derive the Edgeworth expansion of

$$K_n(x,P) = P\{\, n^{1/2}(T(P_n) - T(P)) \,/\, S_n \le x \,\} \; .$$

To explain the form of our result we give the main arguments of the proof which is deferred in chapter V.

We first notice that S_n may be approximated by $\tilde{S}(P_n)$, where $\tilde{S}(Q)$ is some second order differentiable functional equal to $S(P)$ at P. The gradients of $\tilde{S}(P)^2$ are given by

$$S^{(1)}(x,P) := T^{(1)}(x,P)^2 - S^2(P) + 2\, E_Z\, T^{(1)}(Z,P)T^{(2)}(x,Z,P)$$

$$S^{(2)}(x,y,P) := 2\, T^{(1)}(x,P)\, T^{(2)}(x,y,P) + 2\, T^{(1)}(y,P)\, T^{(2)}(x,y,P)$$

$$- 2\, E_Z\, T^{(1)}(Z,P)\, T^{(2)}(Z,x,P) - 2\, E_Z T^{(1)}(Z,P)\, T^{(2)}(Z,y,P)$$

$$+ 2\, E_Z\, T^{(2)}(Z,x,P)T^{(2)}(Z,y,P) + 2\, E_Z\, T^{(1)}(Z,P)T^{(3)}(Z,x,y,P))$$

$$- T^{(1)}(x,P)\, T^{(1)}(y,P)\ ,$$

where Z has distribution P . $S^{(1)}(X_i,P)$ and $S^{(2)}(X_i,X_j,P)$ are no more than the first and second order influence functions of S_n^2 (recall Theorem I.3.2 for the form of $T^{(1)}(.,P)$).

Then to obtain the Edgeworth expansion of $n^{1/2}(T(P_n) - T(P)) / \tilde{S}(P_n)$ notice that we have

$$n^{1/2}(T(P_n) - T(P)) / \tilde{S}(P_n) = n^{1/2}(f(P_n)-f(P))$$

where

$$f(Q) := (T(Q)-T(P))/\tilde{S}(Q)$$

But $f(Q)$ is differentiable and has gradients equal to

$$f^{(1)}(X_i,P) = T^{(1)}(X_i,P)/S(P)\ ,$$

$$f^{(2)}(X_i,X_j,P) = T^{(2)}(X_i,X_j,P)/S(P) - (\ T^{(1)}(X_i,P)\ S^{(1)}(X_j,P) +$$

$$T^{(1)}(X_j,P)\ S^{(1)}(X_i,P)\ )/(2S(P)^3)\ ,$$

$$f^{(3)}(X_i,X_j,X_k,P) = T^{(3)}(X_i,X_j,X_k,P)/S(P)$$

$$- \Sigma_{\sigma\in\mathfrak{S}_3}\ T^{(2)}(X_{\sigma(i)},X_{\sigma(j)},P)\ S^{(1)}(X_{\sigma(k)},P)\ /\ 2S(P)^3$$

$$- \Sigma_{\sigma\in\mathfrak{S}_3} \; S^{(2)}(X_{\sigma(i)},X_{\sigma(j)},P)\; T^{(1)}(X_{\sigma(k)},P) \,/\, 2S(P)^3$$

$$+ 3\,(\Sigma_{\sigma\in\mathfrak{S}_3} \; S^{(1)}(X_{\sigma(i)},P)\; S^{(1)}(X_{\sigma(j)},P)\; T^{(1)}(X_{\sigma(k)},P)/4S(P)^5\;),$$

where $\mathfrak{S}_3$ is the group of permutations of 3 elements.
The Edgeworth expansion of $n^{1/2}(T(P_n) - T(P))/S_n$ may then be deduced from (4.2). We have

(5.9) $$K_n(x,P) = P\{\, n^{1/2}(T(P_n) - T(P)) \,/\, S_n \le x \,\}$$

$$= \mathcal{E}_{2,K}(x,P) + o(n^{-1}) ,$$

$$\mathcal{E}_{2,K}(x,P) := \Phi(x) - n^{-1/2}\left(\frac{K'_{1,1}(P)}{2} + \frac{K'_{1,3}(P)}{6}\,(x^2-1)\right) \phi(x)$$

$$- n^{-1}\left(x\,\frac{K'_{2,2}(P)-K^{Y}_{2,2,n}(P)+\Delta_{Y,n}}{2} + \frac{K'_{2,4}(P)}{24}\,(x^3-3x) + \frac{K'_{2,6}(P)}{72}\,(x^5 - 10x^3 - 15x) \right) \phi(x)$$

where

(5.9′) $$\Delta_{Y,n} = 2\sigma_{Y,n}(\sigma_{Y,n} - \beta_{Y,n}) + \sigma_{Y,n}^{\;2} - 1 ,$$

and the coefficients $K'_{i,j}(P)$ are the coefficients $K_{i,j}(P)$ defined in (4.3) when we replace $T^{(1)}(X_i,P)$, $T^{(2)}(X_i,X_j,P)$, $T^{(3)}(X_i,X_j,X_k,P)$ respectively by $f^{(1)}(X_i,P)$, $f^{(2)}(X_i,X_j,P)$ and $f^{(3)}(X_i,X_j,X_k,P)$.

If the variance $S^2(P)$ is known, as assumed in Hall (1986) and Haeusler (1992), the Edgeworth expansion up to $o(n^{-1})$ for

$$H_n(x,P) = P\{\, n^{1/2}(T(P_n) - T(P))/S(P) \le x \,\}$$

is given in section I.3.

The last step is to plug $\gamma_{W,n}(\alpha)$ into $H_n(.,P)$ and $K_n(.,P)$ to get the coverage probabilities

(5.10)
$$p^H_{c,n}(\alpha) := P\{\, n^{1/2}(T(P_n) - T(P))/S(P) \le \gamma_{W,n}(\alpha) \,\}$$

$$= \alpha + n^{-1/2}\left(\frac{\Delta K^{Y,n}_{1,1}(P)}{2} + \frac{\Delta K^{Y,n}_{1,3}(P)}{6} (u_\alpha^{\,2}-1) \right) \phi(u_\alpha)$$

$$- n^{-1} \frac{K_{2,2}(P)}{2} u_\alpha + n^{-1} \left(\frac{\Delta K^{Y,n}_{2,4}(P)}{24} (u_\alpha^{\,3}-3\, u_\alpha) \right.$$

$$\left. + \frac{\Delta K^{Y,n}_{2,6}(P)}{72} (u_\alpha^{\,5}-10u_\alpha^{\,3}+15u_\alpha) \right) \phi(u_\alpha)$$

$$+ n^{-1} \left(\frac{\Delta K^{Y,n}_{1,3}(P)}{3} - \frac{\Delta K^{Y,n}_{1,1}(P)}{4} - \frac{\Delta K^{Y,n}_{1,3}(P)}{12} (u_\alpha^2 - 1) \right)$$

$$\left(\frac{K^Y_{1,1,n}(P)}{2} + \frac{K^Y_{1,3,n}(P)}{6} (u_\alpha^{\,2}-1) \right) u_\alpha\, \phi(u_\alpha)$$

$$+ n^{-1}\, S(P)^{-1}\, E_P T^{(1)}(X,P)\, \Pi_2^{Y,n(2)}(X,P)\, u_\alpha\, \phi(u_\alpha) + o(n^{-1})$$

as $n \to \infty$, with the following notation

$$\Delta K^{Y,n}_{i,j}(P) = K^{Y,n}_{i,j}(P) - K_{i,j}(P) ,$$

There, the coefficients $K_{i,j}(P)$ are defined as in section I.3 and the $K^{Y,n}_{i,j}(P)$ are the terms $K^Y_{i,j,n}(P)$ defined in (4.9) in which we take the limit as $n \to \infty$ of the $\overline{\tau_1}$, $\overline{\tau_{1,1}}$ and so on. More precisely we have

$$K^{Y,n}_{1,1}(P) := \sigma_{Y,n}\, E_P T^{(2)}(X,X,P)/S(P) ,$$

$$K^{Y,n}_{1,3}(P) := \beta_{Y,n}\, E_P\, T^{(1)}(X,P)^3/S(P)^3$$

$$+ 3\, \sigma_{Y,n}\, E_P T^{(1)}(X,P) T^{(1)}(Y,P) T^{(2)}(X,Y,P)/S(P)^3 ,$$

$$K^{Y,n}_{2,2}(P) := 4^{-1} \sigma^2_{Y,n} \left(E_P T^{(2)}(X,X,P)/S(P) \right)^2$$

$$+ 2^{-1} \sigma^2_{Y,n} \left(E_P T^{(2)}(X,Y,P)^2/S(P)^2 + 2 \right)$$

$$+ \beta_{Y,n}\, \sigma_{Y,n}\, E_P T^{(1)}(X,P) T^{(2)}(X,X,P)/S(P)^2$$

$$+ \sigma_{Y,n}^{2} E_P T^{(1)}(X,P)T^{(3)}(X,Y,Y,P)/S(P)^2 + \Delta_{Y,n} + 1 ,$$

$$K_{2,4}^{Y,n}(P) := 2 K_{1,1}^{Y,n}(P) K_{1,3}^{Y,n}(P) + K_{Y,n} E_P T^{(1)}(X,P)^4/S(P)^4 - 3$$

$$+ 12 \beta_{Y,n} \sigma_{Y,n} \left(E_P T^{(1)}(X,P)T^{(1)}(Y,P)^2 T^{(2)}(X,Y,P)/S(P)^4 - 1 \right)$$

$$+ 12 \sigma_{Y,n}^{2} \left(E_P T^{(1)}(X,P)T^{(1)}(Y,P)T^{(2)}(X,Z,P)T^{(2)}(Y,Z,P)/S(P)^4 + 1\right)$$

$$+ 4 \sigma_{Y,n}^{2} E_P T^{(1)}(X,P)T^{(1)}(Y,P)T^{(1)}(Z,P)T^{(3)}(X,Y,Z,P)/S(P)^4 ,$$

$$K_{2,6}^{Y,n}(P) := K_{1,3}^{Y,n}(P)^2$$

and

$$\Pi_2^{Y,n(1)}(x,P) := 2^{-1} K_{1,1}^{Y,n(1)}(x,P) + 6^{-1} K_{1,3}^{Y,n(1)}(x,P) (u_\alpha - 1)^2 ,$$

where $K_{1,1}^{Y,n(1)}(x,P)$ and $K_{1,3}^{Y,n(1)}(x,P)$ are respectively the influence function of $K_{1,1}^{Y,n}(P)$ and $K_{1,3}^{Y,n}(P)$, given by

$$K_{1,1}^{Y,n(1)}(x,P) := \sigma_{Y,n}\Big(\big(T^{(2)}(x,x,P) - E_P T^{(2)}(X,X,P)$$

$$+ E_P T^{(3)}(X,X,x,P) + 2 T^{(1)}(x,P) \big)/S(P)$$

$$- S^{(1)}(x,P) E_P T^{(2)}(X,X,P)/(2S(P)^3) \Big) ,$$

$$K_{1,3}^{Y,n(1)}(x,P) := 3 \sigma_{Y,n} \big(T^{(1)}(x,P) E_P T^{(1)}(Y,P)T^{(2)}(x,Y,P)$$

$$- E_P T^{(1)}(X,P)T^{(1)}(Y,P)T^{(2)}(X,Y,P)$$

$$+ 2 E_P T^{(2)}(X,x,P)T^{(1)}(Y,P)T^{(2)}(X,Y,P)$$

$$+ E_P T^{(1)}(X,P)T^{(1)}(Y,P)T^{(3)}(x,X,Y,P) \big)/S(P)^3$$

$$+ \beta_{Y,n} \big(T^{(1)}(x,P)^3 - E_P T^{(1)}(X,P)^3 - 3 S(P)^2 T^{(1)}(x,P)$$

$$+ 3 E_P T^{(1)}(X,P)^2 T^{(2)}(X,x,P) \big)/S(P)^3$$

$$- (3/2) K_{1,3}^{Y,n}(P) S^{(1)}(x,P)/S(P)^2 .$$

Finally we can obtain the analogous coverage probability result for the studentized version.

(5.11)
$$p^K_{c,n} := P\{ n^{1/2}(T(P_n) - T(P))/ S_n \le \gamma_{n,W}(\alpha) \}$$

$$= \alpha + n^{-1/2} \left(\frac{\Delta K'^{Y,n}_{1,1}(P)}{2} + \frac{\Delta K'^{Y,n}_{1,3}(P)}{6} (u_\alpha^{\,2}-1) \right) \phi(u_\alpha)$$

$$+ n^{-1} \left(\frac{\Delta K'^{Y,n}_{2,2}(P)-\Delta_{Y,n}}{2} u_\alpha + \frac{\Delta K'^{Y,n}_{2,4}(P)}{24} (u_\alpha^{\,3}-3u_\alpha) \right.$$

$$\left. + \frac{\Delta K'^{Y,n}_{2,6}(P)}{72} (u_\alpha^{\,5}-10u_\alpha^{\,3}+15u_\alpha) \right) \phi(u_\alpha)$$

$$+ n^{-1} \left(\frac{\Delta K'^{Y,n}_{1,3}(P)}{3} - \frac{\Delta K'^{Y,n}_{1,1}(P)}{4} - \frac{\Delta K'^{Y,n}_{1,3}(P)}{12} (u_\alpha^2 - 1) \right)$$

$$\left(\frac{K^{Y,n}_{1,1}(P)}{2} + \frac{K^{Y,n}_{1,3}(P)}{6} (u_\alpha^{\,2} - 1) \right) u_\alpha \, \phi(u_\alpha)$$

$$+ n^{-1} S(P)^{-1} E_P T^{(1)}(X,P) \, \Pi_2^{Y,n(1)}(X,P) \, u_\alpha \, \phi(u_\alpha) + o(n^{-1})$$

as $n \to \infty$, where

$$\Delta K'^{Y,n}_{i,j}(P) := K^{Y,n}_{i,j}(P) - K'_{i,j}(P) \quad ,$$

$i = 1 , 2$ and $j = 1 , 2 , 3 , 4 , 6$ (see after (5.9) the definition of $K'_{i,j}(P)$).

5.2) Choosing the weights via minimization of the coverage probability error.

Consider now, weights such that

(5.12)
$$\sigma_{Y,n} = 1 + n^{-1/2} g \, (1 + o(1)) \quad \text{as} \quad n \to \infty \quad \text{and}$$

$$\beta_{Y,n} = 1 + n^{-1/2} h \, (1 + o(1)) \quad \text{as} \quad n \to \infty .$$

Then, by (5.11) we get a coverage probability

$$p^{H}_{c,n}(\alpha) = \alpha + n^{-1}\Big((g/2)\ (E_P T^{(2)}(X,X,P)/S(P))$$
$$+ 6^{-1}(h\ E_P T^{(1)}(X,P)^3/S^3(P))\ (u_\alpha^{\ 2}-1)$$
$$+ (3g\ E_P T^{(1)}(X,P)T^{(1)}(Y,P)T^{(2)}(X,Y,P)(u_\alpha^{\ 2}-1)/S(P)^3\Big)\ \phi(u_\alpha)$$
$$+ n^{-1}\Big(-2^{-1}\ K_{2,2}(P)\ u_\alpha$$
$$+ 24^{-1}((K_{Y,n}-1)\ E_P T^{(1)}(X,P)^4/S(P)^4)(u_\alpha^{\ 3}-3u_\alpha)\Big)\phi(u_\alpha)$$
$$+ o(n^{-1}) \quad \text{as} \quad n\to\infty .$$

Therefore, if

(5.13) $$E_P T^{(1)}(X,P)^3 \neq 0 \quad \text{or}$$

(5.14) $$6\ E_P T^{(1)}(X,P)T^{(1)}(Y,P)T^{(2)}(X,Y,P)(u_\alpha^{\ 2}-1) + E_P T^{(2)}(X,X,P)\ S(P)^2 \neq 0$$

we can always find h and g such that

(5.15) $$p^{H}_{c,n}(\alpha) = \alpha + o(n^{-1}) .$$

If (5.13) and (5.14) do not hold, and if

(5.16) $$K_{2,2}(P)/(u_\alpha^{\ 2}-3) > 0$$

we can choose

$$K_{Y,n} = 1 + 12\ K_{2,2}(P)\ S(P)^4/((u_\alpha^{\ 2}-3)E_P T^{(1)}(X,P)^4)$$

to get (5.15) again.

Finally, if neither (5.13), (5.14) nor (5.16) hold, we choose

$$K_{Y,n} = 1 + o(1)$$

such that $p^H_{c,n}(\alpha)$ is as accurate as possible. In this case, we have

$$p^H_{c,n}(\alpha) = \alpha - n^{-1}K_{2,2}(P)u_\alpha \phi(u_\alpha)/2 .$$

Since (5.16) does not hold in this last case, we see that $p^H_{c,n}(\alpha) > \alpha$ so that the bootstrap confidence interval is conservative. But one can also use an estimator of $K_{2,2}(P)$. It should be noted that we have (see the proof of (5.6) in chapter V, section 4) :

$$S_n / \sigma_{\tau,n} = 1 + (1/2n)\, K_{2,2}(P) + n^{-1} + o(n^{-1})$$

When S(P) is unknown, we have under (5.12)

$$p^K_{c,n}(\alpha) = \alpha + n^{-1/2}\Big(2^{-1}\, \Delta K'^{\,Y,n}_{1,1}(P) + 6^{-1}\, \Delta K'^{\,Y,n}_{1,3}(P)\, (u_\alpha{}^2 - 1)\Big)\, \phi(u_\alpha)$$
$$+ o(n^{-1/2})$$

with

$$\Delta K'^{\,Y,n}_{1,1}(P) = \sigma_{Y,n} \Big(\, E\; T^{(2)}(X,X,P)/S(P)\, \Big)$$
$$- \Big(\, E_P T^{(2)}(X,X,P)/S(P) - E_P T^{(1)}(X,P)^3/S(P)^3$$
$$- 2E_P T^{(2)}(X,Y,P)T^{(1)}(X,P)T^{(1)}(Y,P)/S(P)^3\, \Big) ,$$

$$\Delta K'^{\,Y,n}_{1,3}(P) = \beta_{Y,n}\; E_P T^{(1)}(X,P)^3/S(P)^3$$
$$+ 3\sigma_{Y,n}\; E_P T^{(2)}(X,Y,P)T^{(1)}(X,P)T^{(1)}(Y,P)/S(P)^3$$
$$+ 2\; E_P T^{(1)}(X,P)^3/S(P)^3$$
$$+ 3\; E_P T^{(2)}(X,Y,P)T^{(1)}(X,P)T^{(1)}(Y,P)/S(P)^3$$

Therefore, we can achieve the first order correctness, i.e.

$$p^K_{c,n}(\alpha) = \alpha + o(n^{-1/2})$$

if one of the following conditions hold

(5.17) $E_P T^{(2)}(X,X,P) \neq 0$ or

(5.18) $E_P T^{(2)}(X,Y,P) T^{(1)}(X,P) T^{(1)}(Y,P) \neq 0$.

If we can get the second order correctness, the term in $n^{-1/2}$ may be taken equal to 0 . Then, one can minimize the term in n^{-1} in (5.11). This may be done easily in practice, in estimating all the coefficients in this term and using standard numerical optimization techniques.

If (5.17) and (5.18) do not hold, then

$$p^K_{\alpha,n}(\alpha) = \alpha$$
$$+ n^{-1/2} S(P)^{-3} E\, T^{(1)}(X,P)^3 \left(2^{-1} + 6^{-1}(\beta_{Y,n}+2)(u_\alpha^{\,2}-1) \right) \phi(u_\alpha)$$
$$+ o(n^{-1/2}) .$$

Then, if the term in $n^{-1/2}$ if $E\, T^{(1)}(X,P)^3 \neq 0$ we can replace $\gamma_{W,n}$ by

$$\gamma_{W,n} - n^{-1/2} S(P)^{-3} E\, T^{(1)}(X,P)^3 \left(2^{-1} + 6^{-1}(\beta_{Y,n}+2)(u_\alpha^{\,2}-1)\right) \phi(u_\alpha)$$

to get the first order correctness.

Of course this result is not really satisfactory. In fact the right method to choose the weights adequately in the studentized case would be to define a weighted bootstrap analogue of S_n . However, if the explicit form is unknown it may be difficult to do so unless we use a computationally expensive double weighted bootstrap. We noticed in the preceeding sections that S_n may be approximated by the functional $\tilde{S}(P_n)$: a weighted bootstrap analogue would be to use $\tilde{S}(P_{W,n})$. The next steps would be to get the Edgeworth expansion of
$P\{ n^{1/2}(T(P_{W,n}) - T(P_n))/ (\tilde{S}(P_{W,n})\sigma_{W,n}) \leq x \mid P_n \}$

using the same techniques as in the preceeding sections and to choose the weights so as to get the analogue of (5.15). Most of the calculations are similar : it suffices to replace $T^{(1)}(.,P)$, $T^{(2)}(.,P)$ and $T^{(3)}(.,P)$ by the gradients of the functional $f(Q)$ i.e. $f^{(1)}(.,P)$, $f^{(2)}(.,P)$ and $f^{(3)}(.,P)$.

5.3 Some simulation results

We now return to the simulations given in I.5.

a) the mean

Recall that $X_1, \ldots, X_n$ are i.i.d. exponentially distributed r.v.'s with mean 1 . $W^{(1)}$ and $W^{(3)}$ are as defined in part I.5 , except that we now require for $W^{(3)}_{i,n}$ that

$$\sigma^2_{W,n} = 1 + o(n^{-1}) \; , \; \beta_{W,n} = 1 + h/\sqrt{n} + o(n^{-1/2}) \; , \; K_{W,n} = k + 3 \tag{5.19}$$

where h and k are chosen such that

$$\hat{\beta}_{X,n}(u_\alpha^2 - 1)(h/(6u_\alpha) - \hat{\beta}_{X,n}/4) + (\hat{K}_{X,n} - 3)(3u_\alpha^2 - 1)/24 \tag{5.20}$$
$$+ \hat{K}_{X,n}\, k\, (u_\alpha^2 - 3)/24 = 0$$

(see sections 3.2 and 5.1 for the justifications).

The sample size is n = 100 . We make nboot = 1000 bootstrap resamplings and the Bootstrap procedure is repeated nrep = 500 times.

We choose two levels $\alpha = 4.5$ and $\alpha = 4.55$ because for $n = 100$ it is always possible to generate weights of type $W^{(3)}$ with h and k satisfying equations (5.19) (5.20). Indeed from a theoretical point of view it is always possible to generate weights of the form (5.19) provided that the sample size n is large enough. But for a fixed n , if we choose k large enough (to ensure that Cauchy-Schwarz inequality with the variance is satisfied) we may get a too large value of h by solving (5.20). This is the case if u_α is far from $\pm\sqrt{3}$ corresponding to the values $\alpha = 4.16\%$ and $\alpha = 95.83\%$ ...

Table 8 compares the quantile and the coverage probability of the true, asymptotic and Bootstrap distributions of the mean standardized by the true variance, for sample size 100. Recall that the adequate weights are not the same for $\alpha = 4.5$ and $\alpha = 4.55$.

	$H_n(.,P)$		$H_n(.,P)$	
n=100	Quant. (α=4.5)	Cover. %	Quant. (α=4.55)	Cover. %
True	-1.630	4.50	-1.627	4.55
Gaussian	-1.695	3.85	-1.690	3.90
$W^{(1)}$	-1.581 (0.11)	5.05 (1.52)	-1.548 (0.19)	5.50 (2.38)
$W^{(3)}$	-1.629 (0.13)	4.51 (1.68)	-1.613 (0.23)	4.61 (2.52)

Table 8: Coverage probability of the (weighted) bootstrap quantile

The simulations are quite interesting : first of all, the results on Efron's Bootstrap (i.e with $W^{(1)}$) are a bit surprising. It is not really better than the asymptotic distribution. In some cases it is even worse! The results with $W^{(3)}$ are very encouraging; we almost have the right level. However the standard deviations of the estimated quantile and of the achieved level (over 500 replications of the Bootstrap procedure) are large. This suggests that nboot = 1000 is not large enough to estimate adequately the quantile with this method. If we admit that the standard deviation is of order K/sqrt(nboot) (see Hall (1986b) for some justifications), the simulations suggest that nboot should be greater than 10 000 000 to obtain an error (due to the resampling procedure) inferior to 1% of the original level.

When the original level goes away from 4.16% , the results deteriorate, due to the fact that, for a fixed n=100 , we are unable to properly generate the weights with h and k solving (5.20).

b) Case of the functional $\int F1\, dF2$

The weights $W^{(3)}$ are constructed as in a) with h and k chosen so that the theoretical coverage probability of the weighted bootstrap quantile equals $\alpha + o(n^{-1})$ (see 5.1). In that case the results (see table 9) are even more impressive. In average, the coverage probability of the weighted Bootstrap is very close to the true level. However the standard deviation suggests one more time that the number of Bootstrap resampling nboot = 1000 is not sufficient to construct a reliable confidence interval.

	$H_n(.,P)$		$H_n(.,P)$	
n=100	Quant. (α=5%)	Cover. %	Quant. (α=4.5%)	Cover. %
True	-1.703	5.0	-1.759	4.5
Gaussian	-1.644	5.4	-1.690	5.0
$W^{(1)}$	-1.753 (0.110)	4.7 (1.62)	-1.765 (0.114)	4.4 (1.86)
$W^{(3)}$	-1.706 (0.139)	4.9 (1.86)	-1.752 (0.128)	4.5 (2.07)

Table 9: Coverage probability of the (weighted) bootstrap quantile, second order and third order correction, n = 100, mean value over 500 replications

II.6) Conditional large deviations

The last criterion that we investigate for choosing the weights is large deviations which are related to Bahadur and Chernoff efficiency (see e.g. Bahadur (1971), Chernoff (1952) and also Serfling (1980) with references therein). Here, we consider the set $\mathcal{P}$ of all p.m.'s on a separable Banach space B endowed with the topology of weak convergence. We shall consider the empirical p.m. of a P i.i.d. distributed B-valued sample $X_1, \ldots, X_n$, i.e.

$$P_n = n^{-1} \Sigma_{i=1}^n \delta_{X_i} ,$$

and its bootstrap analogue

$$P_{W,n} = \Sigma_{i=1}^n W_{i,n} \delta_{X_i} . \tag{6.1}$$

One can require that $P_{W,n}$ mimics P_n in the large deviation sense, i.e. that for a reasonable class of sets Ω of p.m., we have

$$\lim_{n\to\infty} n^{-1} \log P\{ P_n \in \Omega \} \tag{6.2}$$

$$= \lim_{n\to\infty} n^{-1} \log P\{ P_{W,n} \in \Omega \mid P_n \} \quad P\text{ - a.s.}$$

A consequence of the results of Groeneboom, Oosterhoff and Ruymgaart (1979) is that, for any weakly closed set $\Omega \in \mathcal{P}$,

$$\limsup_{n\to\infty} n^{-1} \log P\{ P_n \in \Omega \} \le - K(\Omega,P)$$

and, for any weakly open $\Omega \in \mathcal{P}$,

$$\liminf_{n\to\infty} n^{-1} \log P\{ P_n \in \Omega \} \ge - K(\Omega,P)$$

where, for any $\Omega \subset \mathcal{P}$,

$$K(\Omega,P) := \inf_{Q\in\Omega} K(Q,P) \quad \text{and}$$

$$K(Q,P) := \begin{cases} \int \log(dQ/dP)\, dQ & \text{if } Q \ll P \\ \infty & \text{otherwise} \end{cases}$$

is the Kullback-Leibler information number between Q and P .

For Efron's bootstrap we rewrite (6.1) as

$$P^*_{m_n} := m_n^{-1} \Sigma_{i=1}^{m_n} \delta_{X_i^*} ,$$

where m_n is the resampling size and $X_1^*, \ldots, X_{m_n}^*$ are i.i.d. P_n . Chaganty and Karandikar (1992) proved the following result.

<u>Theorem 6.1</u> : **Whenever** $\lim_{n\to\infty} m_n = \infty$ **, we have for any closed subset** $\Omega \subset \mathcal{F}$

$$\limsup_{n\to\infty} m_n^{-1} \log P\{ P^*_{m_n} \in \Omega \mid P_n \} \le - K(\Omega,P) \qquad P\text{ - a.s.}$$

and for any open subset $\Omega \subset \mathcal{F}$,

$$\liminf_{n\to\infty} m_n^{-1} \log P\{ P^*_{m_n} \in \Omega \mid P_n \} \ge - K(\Omega,P) \qquad P\text{ - a.s.}$$

Therefore, we see that Efron's bootstrap with resampling size m_n mimics the behaviour of the empirical d.f. as far as large deviations are concerned, and we deduce from results of Groeneboom, Oosterhoff and Ruymgaart (1979) that (6.2) holds for instance on convex sets Ω .

For the weighted bootstrap, we investigate the case where the weights are i.i.d. $\mathcal{Y}$. Thus, we denote

$$P_{Y,n} := n^{-1} \Sigma_{i=1}^{n} Y_{i,n} \delta_{X_i}$$

the bootstrapped d.f. We shall assume that the moment generating function

(6.3) $$L(t) := \int_{\mathbb{R}} \exp(tx)\, d\mathcal{Y}(x) \quad \text{is finite in a neighborhood of } 0 ,$$

and define the Chernoff function

$$\gamma(x) := \sup\{ tx - \log L(t) : t > 0 \} .$$

Though it is not really necessary, it is convenient (and will not really modify our basic conclusion) to assume further that

(6.4) $$L(.) \text{ is defined on } (-\infty , t_+) \text{ for } t_+ := \sup\{ t : L(t)<\infty \} > 0 ,$$

and

(6.5) $$m(t) = d(\log L(t))/dt \text{ is an injection into } \mathbb{R}^+ .$$

The precise meaning of (6.5) is that $m^{-1}(.)$ is defined on $\mathbb{R}^+$.

Theorem 6.2 : **If the weights $Y_{i,n}$ are i.i.d. $\mathcal{Y}$, then under** (6.3)-(6.5) **we have, for any closed $\Omega \in \mathcal{P}$,**

$$\limsup_{n\to\infty} n^{-1} \log P\{ P_{Y,n} \in \Omega \mid P_n \} \le - J(\Omega,P) \quad P \text{ - a.s.}$$

and, for any open subset $\Omega \subset \mathcal{P}$,

$$\liminf_{n\to\infty} n^{-1} \log P\{ P_{Y,n} \in \Omega \mid P_n \} \ge - J(\Omega,P) \quad P \text{ - a.s.}$$

with

$$J(\Omega,P) := \inf_{Q\in\Omega} J(Q,P) \quad \textbf{and}$$

(6.6) $$J(Q,P) = \begin{cases} \int \gamma(dQ/dP)\, dP & \textbf{if } Q \ll P \\ \infty & \textbf{otherwise.} \end{cases}$$

Remark 6.1 : A careful look at the proof of Theorem 6.2 shows that it can be extended when the weights $Y_{i,n}$ are i.i.d. $\mathcal{Y}_n$, even if $\mathcal{Y}_n$ depends on the data $X_1, \ldots, X_n$, provided that $\mathcal{Y}_n$ converges a.s. to some $\mathcal{Y}$ as $n \to \infty$. A slight modification of the proof also enables one to consider the case of iid-generated weights $nW_{i,n} = Y_{i,n}/\overline{Y}_n$ where the $Y_{i,n}$ are i.i.d. $\mathcal{Y}_n$ and $\overline{Y}_n = n^{-1} \Sigma_{i=1}^{n} Y_{i,n}$. However, in all those cases, the following conclusion remains unchanged :

Observe that we can rewrite

$$K(Q,P) = \int (dQ/dP) \log(dQ/dP)\, dP$$

so that $J(.,P) = K(.,P)$ implies

$$\gamma(x) = x \log x \ . \tag{6.7}$$

Since $\log L$ is the conjugate of γ (see e.g. Rockafellar (1970)), (6.7) yields $\log L(x) = \exp(x-1)$ which is not the logarithm of a Laplace transform of a probability d.f. since $L(0) = \exp(1/e)$. Therefore, (6.2) cannot hold for any convex set $\Omega \subset \mathcal{F}$.

Now, we can weaken a little bit the requirement $J(.,F) = K(.,F)$ and just ask that

$$J(.,P) = K(.,P) + c \tag{6.8}$$

for some constant c .

Then, we get that $\gamma(x) = x \log x + c$ and

$$\log L(x) = \exp(x-1) + c \ .$$

Since we must have $\log L(0) = 0$, we must have $c = -1/e$. Then, $L(x) = \exp((e^x-1)/e)$ is the Laplace transform of the Poisson(1) distribution. Thus, we have the following result.

Theorem 6.3 : **Under the assumptions of Theorem 6.2,** (6.8) **holds iff** $\mathcal{Y}$ **is the** Poisson(1) **distribution. In this case** $c = -1/e$.

The point, here, is that if we consider Efron's bootstrap, the multinomial $\mathcal{M}(n,1/n,\ldots,1/n)$ behaves asymptotically as a vector of i.i.d. Poisson(1) r.v.'s. Thus, Theorems 6.2 and 6.3 essentially say that we cannot do anything really different from Efron's bootstrap if we want to mimic properly the large deviations of P_n . This conclusion remains unchanged if we consider weights of the form $Y_{i,n}/\bar{Y}_n$.

II.7) Conclusion

This chapter clearly shows that the choice of the adequate weights depend on the priorities of the statistician. From the point of view of the adequacy of the whole distribution, we can do as well as the usual bootstrap for a very large class of weights, but this does not really improve Efron's Bootstrap. From the point of view of coverage probability, we can do always better provided that we choose the first four moments of the weights as a function of the statistics and the sample. Practically it is not always possible to generate the adequate weights but, when it is possible we obtain a very important correction and very accurate confidence intervals. Finally from a large deviation aspect, Efron's Bootstrap seems to be the most interesting method.

CHAPTER III

SPECIAL FORMS OF THE BOOTSTRAP

1) Introduction

In chapter II, we discussed the choices of the weights, when one wants to bootstrap regular functionals as described in the first chapter. However one can find functionals which do not satisfy the assumptions of chapter I or a sample which is not i.i.d. but is obtained from i.i.d. r.v.'s. The aim of this chapter is to investigate three of such situations : what can we do if we want to bootstrap an empirical process when the parameters are estimated ? How can the extreme values be bootstrapped ? What happens to the bootstrap of the mean when the limiting distribution is non gaussian ?

It turns out that the general answer is that the bootstrap must be modified, adding more randomness than usually to mimic extra fluctuations.

In this chapter, we consider real valued observations, and we apply the weighted approximation and quantile transform technique (see e.g. the volume edited by Hahn, Mason and Weiner (1991)). In the first section we develop a weigthed approximation of empirical process with estimated parameters (Theorem 2.1) and its generalized bootstrap version (Theorem 2.2).

The natural companion to this technique is the extreme value theory since it is the usual tool that one needs when the weighted approximation fails to give limiting distribution of the statistic (see e.g. M. Csörgo, S. Csörgo, Horváth and Mason (1986b) for this approach applied to the mean). Hence, in the second sub-section we shall investigate the generalized bootstrap of the extreme values. The main result is that the usual bootstrap with an undersampling size is somehow the only one which is consistent for the extreme value problems (Theorems 3.1 and 3.2).

2) Bootstrapping an empirical d.f. when parameters are estimated or under local alternatives

In this subsection we are concerned with the following type of problem. Assume that we observe a sequence $X_1, \ldots, X_n$ of i.i.d. real valued r.v.'s with common d.f. F , and on the basis of $X_1, \ldots, X_n$, we want to test whether F is gaussian. If the mean $m = E\,X_i$ and the variance $\sigma^2 :=$ $\mathrm{Var}(X_i)$ are known, we can test whether $\Phi((X_i-m)/\sigma)$ are i.i.d. with uniform distribution d.f. $\mathcal{U}_{[0,1]}$ (where Φ is the $\mathcal{N}(0,1)$ c.d.f.) and use the empirical d.f.

$$(2.1) \qquad n^{-1} \Sigma_{i=1}^{n} \mathbb{I}\{ \Phi((X_i - m)/\sigma) \le x \} , \quad 0 \le x \le 1 .$$

But since generally m and σ^2 are unknown, we replace them by some estimators μ_n and σ_n^2 based on $X_1, \ldots, X_n$ and we consider the empirical d.f. with estimated parameters

$$n^{-1} \Sigma_{i=1}^{n} \mathbb{I}\{ \Phi((X_i-\mu_n)/\sigma_n) \le x \} , \quad 0 \le x \le 1 .$$

Asymptotic theory for empirical process with estimated parameters have been investigated by Durbin (1973a, 1973b). More work has been done through strong approximation techniques by Bürke and M. Csörgő (1976) and Bürke, M. Csörgő, S. Csörgő and Révész (1979) (see also Csörgő and Révész (1981)). The Efron bootstrap analogue has been given in Bürke and Gombay (1988).

We can rewrite (2.1) as

$$(2.2) \qquad n^{-1} \Sigma_{i=1}^{n} \mathbb{I} \{ U_i \le \Phi((\sigma_n/\sigma)\, \Phi^{-1}(x) + (\mu_n-m)/\sigma) \} , \ 0 \le x \le 1 .$$

In view of (2.2) it is natural to introduce the perturbated empirical d.f. of a uniform $\mathcal{U}_{[0,1]}$ sample , $U_1, U_2, \ldots, U_n$.

$$(2.3) \qquad \mathbb{D}_n(x) := n^{-1} \Sigma_{i=1}^{n} \mathbb{I}\{ U_i \le g_n(x) \}$$

where $(g_n)_{n\ge 1}$ is a sequence of possibly random functions from $[0,1]$ to $[0,1]$, such that in some sense $\lim_{n\to\infty} g_n = I$, the identity function.

The function $\mathbb{D}_n$ defined in (2.3) may be also useful to describe the behaviour of the empirical process under sequences of local alternatives. For example, if $X_1^{(n)}, \ldots, X_n^{(n)}$ are i.i.d $N(\mu_n,\sigma_n^2)$ then we have

$$(2.4) \qquad n^{-1} \Sigma_{i=1}^{n} \mathbb{I}\{ X_i^{(n)} \le \Phi^{-1}(x) \}$$

$$= n^{-1} \Sigma_{i=1}^{n} \mathbb{I}\{ U_i \le \Phi((\Phi^{-1}(x)-\mu_n)/\sigma_n)) \} , \quad 0 \le x \le 1$$

Thus if $\mu_n \longrightarrow 0$ and $\sigma_n \longrightarrow 1$ as $n \to \infty$, (2.4) is of the form (2.3).

Now assume we base our test on $\mathbb{D}_n$. How can we bootstrap this test ? One method to investigate this problem is to use the weighted approximation technique of M. Csörgo, S. Csörgo, Horváth and Mason (1986a,b) and its bootstrap analogue by S. Csörgo and Mason (1989) (see Einmahl and Mason (1991) for the generalized bootstrap analogue).

Our first theorem in this subsection provides sufficient conditions on g_n for a weighted approximation for $\mathbb{D}_n$ to hold. We assume that there exists a deterministic sequence of functions $\tilde{g}_n$ such that

(2.5) $$\sup_{t\in[1/n,1-1/n]} \sup_{|s|\le 1/(2n)} n^{1/2} \frac{|g_n(t+s)-\tilde{g}_n(t)|}{(t(1-t))^{1/2}} = O_p(1)$$

as $n \to \infty$. If g_n is deterministic then we can of course take $\tilde{g}_n = g_n$ so that (2.5) is always true.

We also assume that we have the linear bound

(2.6) $$\| g_n(1-g_n)/(I(1-I)) \|_n = O_p(1) ,$$

where $\| \; \|_n$ denotes the semi-norm defined by

$$\| f \|_n = \sup_{t\in[1/n,1-1/n]} |f(x)| .$$

We also suppose that the sequence $U_1, U_2, \dots, U_n$ is defined on the probability space of M. Csörgo, S. Csörgo, Horváth and Mason (1986a) or Mason and Van Zwet (1987). $(B_n)_{n\ge 1}$ denotes the sequences of Brownian bridge of M. Csörgo, S. Csörgo, Horváth and Mason (1986a).

<u>Theorem 2.1</u> : **Under (2.5) and (2.6), for any** $0 \le \nu < 1/4$ **we have**

(2.7) $$n^{\nu} \left\| \frac{n^{1/2}(\mathbb{D}_n - \tilde{g}_n) - B_n \circ \tilde{g}_n + n^{1/2}(\tilde{g}_n - g_n)}{(I(1-I))^{1/2-\nu}} \right\|_n = O_p(1) \quad \textbf{as} \quad n \to \infty .$$

It is much more complicated to find nice conditions to obtain a weighted approximation of the quantile process $n^{1/2}(\mathbb{D}_n^{\leftarrow}(x) - \tilde{g}_n^{\leftarrow}(x))$ where

$$\mathbb{D}_n^{\leftarrow}(x) := \inf \{ t : \mathbb{D}_n(t) \ge x \} .$$

It seems that one needs to know more on the form of g_n to obtain an approximation. However we shall see hereafter that one can often follow the proof of the result for the renewal spacings quantile process in Barbe (1993), writing that $\mathbb{D}_n^{\leftarrow}(x) \overset{d}{=} g_n^{\leftarrow} \circ \mathbb{U}_n^{\leftarrow}$, where $\mathbb{U}_n^{\leftarrow}$ is the empirical quantile distribution function of a uniform $\mathcal{U}_{[0,1]}$ sample of size n and then using the weighted approximation of M. Csörgo, S. Csörgo, Horváth and Mason (1986a).

Similarly to the spacing process, if we use

$$\Sigma_{i=1}^{n} W_{i,n} \, \mathbb{I}\{ U_i \leq g_n(x) \}$$

as a bootstrapped version of $\mathbb{D}_n$ we cannot get back to the correct limiting process in (2.7). To mimic the fluctuation due to g_n in (2.7) we define

$$\mathbb{D}_{W,n} := \Sigma_{i=1}^{n} W_{i,n} \, \mathbb{I}\{ U_i \leq g_{W,n}(x)\}$$

where the $W_{i,n}$'s are as previously the exchangeable weights and $g_{W,n}(.)$ is a function depending on the same weights $W_{i,n}$'s and which satisfies some assumptions listed hereafter.

We assume that the weights satisfy the Einmahl and Mason (1992) conditions (see Appendix 1), namely, that

(2.8) $$\Sigma_{i=1}^{n} W_{i,n} = 1$$

(2.9) $$n \, \Sigma_{i=1}^{n}(W_{i,n} - n^{-1})^2 = \sigma^2 + O_p(n^{-1/2})$$

(2.10) $$\limsup_{n\to\infty} E(nW_{i,n} - 1)^4 < \infty$$

If the probability space where the weights live is rich enough (possibly enlarged) we can assume that there exists a fixed Brownian bridge B such that Theorem 2.1′ of Einmahl and Mason (1992) holds, that is

(2.11) $$n^{\nu} \left\| \frac{n^{1/2} \, \Sigma_{i=1}^{[nt]}(W_{i,n} - t) - \sigma B(t)}{(t(1-t))^{1/2-\nu}} \right\|_n = O_p(1) \quad \text{as } n\to\infty .$$

for any $0 \leq \nu < 1/4$.

We shall assume that there exists a functional γ such that

$$n^{\nu} \left\| \frac{n^{1/2}(\tilde{g}_n - g_n) - \gamma(B_n)}{(I(1-I))^{1/2-\nu}} \right\|_n = O_p(1) \quad \text{as} \quad n \to \infty$$

and that this is mimiced by $g_{W,n}$ in the sense that

$$(2.13) \qquad n^{1/2} \left\| \frac{n^{1/2}(g_n - g_{W,n}) - \sigma\gamma(B_n)}{(I(1-I))^{1/2-\nu}} \right\|_n = O_p(1) \quad \text{as} \quad n \to \infty$$

and that the behaviour of $g_{W,n} - g_n$ is controlled by

$$(2.14) \qquad n^{1/2} \left\| \frac{(g_n - g_{W,n})}{(I(1-I))^{1/2}} \right\|_n = O_p(1) \quad \text{as} \quad n \to \infty .$$

Theorem 2.2 : **Under** (2.8)-(2.14) **and assumptions of Theorem 2.1,**

$$(2.15) \qquad n^{\nu} \left\| \frac{n^{1/2}(\mathbb{D}_{W,n} - D_n) - \sigma\, B_n \circ \tilde{g}_n + \sigma\gamma(B_n)}{(I(1-I))^{1/2-\nu}} \right\|_n = O_p(1) ,$$

for any fixed $0 \le \nu \le 1/4$.

To illustrate the use of Theorems 2.1 and 2.2 and to show how to deal with the quantile function, we now investigate the location scale model.

Let $F((x-m)/\sigma)$ a c.d.f. on $\mathbb{R}$ with location parameter m and scale parameter $\sigma > 0$. Let $(X_i)_{i\ge1}$ be an i.i.d. sequence of r.v.'s with common distribution function F . For the empirical process with estimated parameter we consider

$$(2.16) \qquad \tilde{F}_n(x) := n^{-1} \Sigma_{i=1}^n \mathbb{I}\{ \sigma_n^{-1}(X_i - \mu_n) \le x \}$$

where μ_n and $\sigma_n > 0$ are respectively some estimators of m and σ .

Under the model with a sequence of local alternatives, we consider (2.16) when μ_n and $\sigma_n > 0$ are deterministic and converge respectively to m and σ at a rate precised below.

We may assume without loss of generality that $X_i = F^{\leftarrow}(U_i)$ where $(U_i)_{i\ge1}$ is an i.i.d. sequence of $\mathcal{U}_{[0,1]}$ r.v.'s. Then, we can rewrite (2.16) under the form (2.3) with

$$g_n(x) := F\big((\sigma_n/\sigma)\, F^{\leftarrow}(x) + ((\mu_n - m)/\sigma) \big) , \; 0 \le x \le 1 .$$

Since μ_n and σ_n converge in some sense (according to their nature, estimated parameters or deterministic sequences) to m and σ it is natural to introduce

$$\tilde{g}_n(x) := x$$

To check (2.5), denote

$$\tilde{\sigma}_n := (\sigma_n - \sigma)/\sigma \quad \text{and} \quad \tilde{\mu}_n := (\mu_n - m)/\sigma \ .$$

Observe that

$$g_n(t) - \tilde{g}_n(t) = F(\, F^{\leftarrow}(t) + \tilde{\sigma}_n F^{\leftarrow}(t) + \tilde{\mu}_n \,) - t$$

$$= \tilde{\sigma}_n F^{\leftarrow}(t)\; f\circ F^{\leftarrow}(t) + \tilde{\mu}_n f\circ F^{\leftarrow}(t)$$

$$+ 1/2\; (\tilde{\sigma}_n F^{\leftarrow}(t) + \tilde{\mu}_n)^2\; f'(F^{\leftarrow}(t)+\theta_n(t))\; (\tilde{\sigma}_n F^{\leftarrow}(t)+\tilde{\mu}_n) \ ,$$

where $\theta_n(.)$ is a function from $[0,1]$ to $[0,1]$. Hence, if

(2.17) $f = F'$ exists and is continuously differentiable except at some isolated points,

we have

(2.18) $$n^{1/2}(\sigma_n - \sigma) = O_p(1) \quad \text{and} \quad n^{1/2}(\mu_n - m) = O_p(1) \ .$$

Moreover, if

(2.19) $$\left\| \frac{2f(x)}{F(x)^{1/2}\,(1-F(x))^{1/2}} \right\| < \infty \quad \text{and}$$

(2.20) $$\|(x^2+1)f'(x)\| < \infty,$$

approximation (2.5) holds, using the same arguments as in Barbe (1993). Condition (2.20) is always satisfied if the density function f is bounded and ultimately concave or convex. Assumption (2.6) is then satisfied.

Generally, if we denote the usual empirical d.f.

$$F_n(x) := n^{-1}\,\Sigma_{i=1}^{n}\,\mathbb{I}\{\,X_i \le x\,\} = n^{-1}\,\Sigma_{i=1}^{n}\,\mathbb{I}\{\,U_i \le F^{\leftarrow}(x)\,\}$$

$$= \mathbb{U}_n \circ F^{\leftarrow}(x)\ ,$$

σ_n and μ_n may be easily expressed as functional of F_n , say

$$\mu_n = M(F_n) \text{ and } s_n = \Sigma(F_n)\ .$$

Clearly if we can linearize Σ and M in a neighborhood of F as described in chapter I, μ_n and σ_n will be asymptotically normal. With $T = \Sigma$ or M , we have

$$n^{1/2}(T(F_n) - T(F)) = \int_0^1 T^{(1)}(F^{\leftarrow}(x),F)\ dB_n(x) + o_p(1) \quad \text{as} \quad n \to \infty\ . \tag{2.21}$$

Let $M^{(1)}(.,F)$ and $\Sigma^{(1)}(.,F)$ be respectively the influence functions of M and Σ at F . Define the gaussian process on $[0,1]$

$$\Gamma(B,x) = B\circ F^{\leftarrow}(x) + \sigma^{-1}\,F^{\leftarrow}(x)\ f\circ F^{\leftarrow}(x) \int_0^1 \Sigma^{(1)}(F^{\leftarrow}(t),F)\ dB(t)$$

$$+\ \sigma^{-1}\ f\circ F^{\leftarrow}(x) \int_0^1 M^{(1)}(F^{\leftarrow}(t),F)\ dB(t)\ ,$$

where B is a brownian bridge. Then Theorem 2.1 provides a weighted approximation of $n^{1/2}(\tilde{F}_n - F)(.)$ by $\Gamma(B_n, F(.))$.

We can obtain a weighted approximation of the quantile process in the same way, noticing that

$$F_n^{\leftarrow} \stackrel{d}{=} g_n^{\leftarrow} \circ \mathbb{U}_n \quad \text{and}$$

$$g_n^{\leftarrow}(x) = F\Big((\sigma/\sigma_n)\ F^{\leftarrow}(x) - (\mu_n - m)/\sigma_n\Big)\ .$$

Assuming that

$$\|\ F^{1/2}(1-F)^{1/2}\ I\ f'/f\ \| < \infty \quad \text{and} \tag{2.22}$$

(2.23) $F^{\leftarrow}(x)f'\circ F^{\leftarrow}(x)/f\circ F^{\leftarrow}(x)$ and its derivative are regularly varying at 0^+ and 1^- ,

$$(2.24)\qquad \int_{\mathbb{R}} (F(x)(1-F(x)))^{1/4}\,dx < \infty ,$$

one can readily infer from the proof of Theorem 2.2 in Barbe (1993) the next result.

Theorem 2.3 : **Under (2.17)-(2.24) the following approximation holds :**

$$\left\| \frac{n^{1/2}(F_n^{\leftarrow}(x) - F^{\leftarrow}(x)) + \Gamma(B_n,x)}{(x(1-x))^{1/2-\nu}} \right\|_n = o_P(1) \quad \textbf{as } n\to\infty .$$

If, instead of having estimated parameters, we are under local alternatives satisfying

$$\lim_{n\to\infty} n^{1/2}(\mu_n - m) = \mathcal{U} \quad\text{and}\quad \lim_{n\to\infty} n^{1/2}(\sigma_n - \sigma) = \mathcal{S} ,$$

it suffices to replace the process Γ by

$$\Gamma_{loc}(x) = B_n\circ F^{\leftarrow}(x) + \mathcal{S}/\sigma\; F^{\leftarrow}(x)\; f\circ F^{\leftarrow}(x) + \mathcal{U}/\sigma\; f\circ F^{\leftarrow}(x) .$$

To bootstrap $\tilde{F}_n$, we have to mimic the fluctuations of g_n . Thus if $\sigma_n =$

$\Sigma(F_n)$ and $\mu_n = M(F_n)$, the bootstrapped d.f. with estimated parameters becomes

$$(2.25)\qquad \tilde{F}_{W,n}(x) := \Sigma_{i=1}^{n} W_{i,n}\; \mathbb{I}\{\, \Sigma(F_{W,n})^{-1}\,(X_i - M(F_{W,n})) \le x \,\} ,$$

where $F_{W,n}$ is the bootstrapped d.f.

$$(2.26)\qquad F_{W,n}(x) := \Sigma_{i=1}^{n} W_{i,n}\; \mathbb{I}\{\, X_i \le x \,\} .$$

Then, Theorem 2.2 provides a weighted approximation of the bootstrap version $F_{W,n}$. It is important to observe that the weights in (2.25) and (2.26) are the same, so that the bootstrapped version may be written

$$(2.27)\qquad \tilde{F}_{W,n}(x) = F_{W,n}(x\Sigma(F_{W,n}) + M(F_{W,n}))$$

If we consider the bootstrap under a sequence of local alternatives, we see that it can be performed only if $\mathcal{S}$ and $\mathcal{U}$ are known, so that they can be plugged into (2.27) instead of $\Sigma(F_{W,n})$ and $M(F_{W,n})$.

Finally we mention that linear bounds on $\tilde{F}_n$ and $\tilde{F}_n^{\leftarrow}$ may also be obtained, using assumptions very similar to Barbe (1993). Therefore the complete tools to apply the weighted approximation techniques are available and a similar metatheorem as the one formulated by S. Csörgő and Mason (1989) holds for the processes studied in this section.

III.3) Bootstrap of the extremes and bootstrap of the mean attracted by a nongaussian stable distribution.

In this section we are interested in bootstrapping the extreme values of a sample. As a by-product, we obtain some informations on the generalized bootstrap of the mean in the case of infinite variance of the underlying distribution.

To explain how the generalized bootstrap can be used in extreme values, we first deal with the uniform case, and then we shall use the quantile transform technique. Therefore, let us consider a sample $U_1, \ldots, U_n$ of i.i.d. r.v.'s which are uniformly distributed over $[0,1]$. Denote $U_{1,n} \le \ldots \le U_{n,n}$ the ordered sample, and introduce $(\omega_i)_{i\ge 1}$ a sequence of i.i.d. r.v.'s exponentially distributed with mean 1 . Then, if we denote

(3.1) $$\Omega_k := \omega_1 + \ldots + \omega_k ,$$

it is well known that for any fixed $k \ge 1$,

(3.2) $$\{ n(1 - U_{n-j,n}) : 1 \le j \le k \} \xrightarrow{d} \{ \Omega_j : 1 \le j \le k \} \quad \text{as} \quad n \to \infty .$$

Now, define a bootstrap sample (in the sense of Efron) , say $U_1^*, \ldots, U_{m_n}^*$ which is i.i.d. with common distribution function the empirical distribution function

$$\mathbb{U}_n(x) := n^{-1}\Sigma_{i=1}^n \mathbb{I}\{ U_i \le x \} .$$

Introduce the order statistics of the bootstrapped sample

$$U_{1,m_n}^* \le \ldots \le U_{m_n,m_n}^* .$$

We shall say that the bootstrap is a.s. (resp. weakly or in probability) consistent if we have for some sequence $(\alpha_n)_{n\geq 1}$

$$P\{ (U_{n,n} - U^{*}_{m_n-j,m_n})/\alpha_n \leq x_j : 1 \leq j \leq k \mid \mathbb{U}_n \}$$

converges a.s. (resp. in probability) to

$$(3.3) \qquad P\{ \Omega_j \leq x_j : 1 \leq j \leq k \}$$

uniformly in $\{ x_j : 1 \leq j \leq k \}$. This means that (3.2) is imitated by the bootstrap.

It is known (Swanepoel (1986), Deheuvels, Mason and Shorack (1992), see also Politis and Romano (1992) for a very general result) that if

$$\lim_{n\to\infty} m_n / n = 0$$

the usual bootstrap is weakly consistent, while if

$$\lim_{n\to\infty} m_n \log\log n / n = 0$$

it is a.s. consistent.

For Efron's bootstrap, it is clear how to bootstrap the extreme values because there is a bootstrapped sample which can be ordered. For the generalized bootstrap, we need another definition.

We rewrite the l.h.s. of (3.2) as

$$\{ n(1-\mathbb{U}_n^{\leftarrow}(1-j/n)) : 1 \leq j \leq k \}$$

where $\mathbb{U}_n^{\leftarrow}$ is the empirical quantile function defined by

$$\mathbb{U}_n^{\leftarrow}(s) := \inf\{ x : \mathbb{U}_n(x) \geq s \} , \ 0 \leq s \leq 1 .$$

We denote

$$\mathbb{U}_{W,n}(x) := \Sigma_{i=1}^{n} W_{i,n} \ \mathbb{I}\{ U_i \leq x \}$$

the bootstrapped d.f. and define the bootstrapped quantile function

$$\mathbb{U}^{\leftarrow}_{W,n}(s) := \inf \{ x : \mathbb{U}_{W,n}(x) \geq s \} .$$

We shall assume that

(3.4) the weights $W_{i,n}$ are non-negative

in order that $\mathbb{U}_{W,n}$ is nondecreasing.

It is natural to look at the bootstrapped extreme values

(3.5) $$U_{W,j,n} := \mathbb{U}^{\leftarrow}_{W,n}(1 - jc_n/n) , \quad (1 \leq j \leq k)$$

where $(c_n)_{n\geq 1}$ is a deterministic sequence of real numbers that we have to choose in a proper way. Then, the bootstrapped distribution function of the k extreme values will be defined as

(3.6) $$P\{ m_n (U_{n,n} - U_{W,j,n}) \leq x_j : 1 \leq j \leq k \mid \mathbb{U}_n \}$$

for some suitable sequence $(m_n)_{n\geq k}$ which may be interpreted as a resampling size. As previously, we say that the bootstrap of the extreme values is a.s. consistent (resp. weakly or in probability) if (3.6) converges a.s. (resp. in probability) to (3.3).

Our next theorem shows that, for the generalized bootstrap of the extremes to be consistent, we need some very special weights, and it also shows how m_n in (3.6) must be related to c_n in (3.5) and to the weights.

Theorem 3.1 : **i) Under (3.4), the generalized bootstrap of the uniform extreme values is consistent in probability if and only if**

(3.7) $$\lim_{n\to\infty} m_n/n = 0 \quad \textbf{and}$$

(3.8) $$t \longmapsto [\, n\, c_n^{-1} \textstyle\sum_{1\leq i\leq nt/m_n} W_{i,n} \,]$$ **converges weakly to a Poisson process, where** $[.]$ **denotes the integer part.**

ii) Moreover, the generalized bootstrap of the extremes is a.s. consistent if

(3.9) $$\lim_{n\to\infty} m_n n^{-1} \log\log n = 0$$ **and (3.7) holds.**

To understand why conditions (3.7) and (3.8) are, in some sense natural, recall that for any $0 < T < \infty$,

$$\{ n(1-\mathbb{U}_n(1-t/n)) : 0 \le t \le T \} \xrightarrow{d} \{ N(t) : 1 \le t \le T \} \tag{3.10}$$

where $N(.)$ denotes a Poisson process. Intuitively, for the bootstrap to work, we need that

$$\{ nc_n^{-1}(1-\mathbb{U}_{W,n}(1-t/m_n)) : 0 \le t \le T \mid \mathbb{U}_n \} \xrightarrow{d} \{ N(t) : 0 \le t \le T \} \tag{3.11}$$

for some sequence $(c_n)_{n\ge1}$ and $(m_n)_{n\ge1}$. Now, if we introduce the exchangeable process

$$\mathbb{W}_n(t) := \Sigma_{1\le i\le nt} W_{i,n} , \tag{3.12}$$

then we have (Einmahl and Mason (1992))

$$\mathbb{U}_{W,n} \overset{d}{=} \mathbb{W}_n \circ \mathbb{U}_n . \tag{3.13}$$

Now, for (3.11) to hold, we need first that $\mathbb{U}_n(.)$ behaves like a continuous function, otherwise $\mathbb{U}_{W,n}(.)$ would have some jumps depending on the sample and the convergence (3.10) would not hold. Of course, when we say 'behaves like a continuous function', this must be understood in the sense that the jumps of $\mathbb{U}_n(.)$, which are of size $1/n$, must be small compared to the local fluctuations of the exchangeable process $\mathbb{W}_n$. This is condition (3.7). Then, $\mathbb{U}_n(.)$ is "smooth" enough, it behaves like the identity function, and for (3.10) to hold, we need $\mathbb{W}(.)$ to be locally Poisson, which is (3.8).

Now, we turn to the general case and consider a sample $X_1 , \ldots, X_n$ of i.i.d. r.v.'s with a common distribution function F , and denote $X_{1,n} \le \ldots \le X_{n,n}$ the ordered sample. We shall assume that the upper extreme value belongs to a domain of attraction, which means that there exists two deterministic sequences $(\alpha_n>0)_{n\ge1}$ and $(\beta_n)_{n\ge1}$ such that

$$(X_{n,n}-\beta_n)/\alpha_n \xrightarrow{d} Z \quad (\text{as } n\to\infty) , \tag{3.14}$$

where Z is a non-degenerate r.v. In this case, we have for any fixed k

(3.15) $$\{ (X_{n-j,n} - Q(1-1/n))/\alpha_n : 1 \le j \le k \} \xrightarrow{d} \{ Z_j : 1 \le j \le k \} ,$$

where

$$\alpha_n = Q(1-n^{-1}) - Q(1-2n^{-1}) \text{ , and}$$

$$Q(u) := \inf\{ x : F(x) \ge u \}$$

is the quantile function, and the Z_j are some non-degenerate r.v. (see Galambos (1978) for an exposition of the classical extreme values theory). From De Haan (1970), we infer that (3.14) holds if and only if

(3.16) $$H_c(x,y,w,z) := \lim_{s \to 0} \frac{Q(1-sx) - Q(1-sy)}{Q(1-sw) - Q(1-sz)}$$

exists and is equal to

$$H_c(x,y,w,z) = (x^{-c}-y^{-c}) / (w^{-c}-z^{-c})$$

with $$H_0(x,y,w,z) = \log(x/y) / \log(w/z) .$$

In this case, we have $Z_j = H_c(\Omega_j, 1, 1, 2)$ where the Ω_j are defined in (3.1). Now, we define the bootstrapped version of (3.15). For this, recall that

$$F_n(x) = n^{-1}\Sigma_{i=1}^n \mathbb{I}(X_i \le x)$$

denotes the empirical distribution function. We need also the empirical quantile function

$$Q_n(s) := \inf\{ x : F(x) \ge s \} , \; 0 \le s \le 1 ,$$

and their generalized bootstrapped analogue

$$F_{W,n}(x) = \Sigma_{i=1}^n W_{i,n} \mathbb{I}\{X_i \le x\} \text{ and}$$

$$Q_{W,n}(u) := \inf\{ x : F_{W,n}(x) \ge u \} .$$

Then the bootstrap analogue of the l.h.s. of (3.15) is

(3.17) $$(Q_{W,n}(1-jc_n/n) - Q_n(1-m_n^{-1}))/\hat{\alpha}_n$$

with

$$\hat{\alpha}_n := Q_n(1-m_n^{-1}) - Q_n(1-2m_n^{-1}) .$$

Theorem 3.2 : **i) If (3.4), (3.7), (3.8) and $\lim_{n\to\infty} m_n = \infty$ hold, then the bootstrap for the extreme values as defined in (3.17) is weakly consistent.**
ii) Moreover, if

$$\lim_{n\to\infty} m_n n^{-1}\log\log n = \infty ,$$

the bootstrap of the extremes is a.s. consistent.

The interesting point in this bootstrap of extremes is that we need some very special weights, such that (3.8) holds or similarly, that the exchangeable process associated with the weights $W_{i,n}$ may be locally approximated by a Poisson process. An example of non-multinomial weights such that this holds may be constructed as follows : let $Y_{i,n}$ $(1\le i\le n)$ be i.i.d. with a Poisson $\mathcal{P}(1)$ distribution. Set $c_n = m_n/n$. Then (3.8) holds with $W_{i,n} = Y_{i,n}/n$.

Now, from our result on extremes, we can easily obtain a result on the bootstrap of the mean in the infinite variance case. From now on, we assume that the distribution function F of our sample belongs to the domain of attraction of a stable distribution $\mathcal{L}_\alpha$ with index $0 < \alpha < 2$. Our next theorem shows how to bootstrap the mean in such a case. We denote $\overline{X}_{W,n} := \Sigma_{i=1}^n W_{i,n}X_i$ the generalized bootstrap mean.

Theorem 3.3 : **If (3.7) and (3.8) hold, then**

$$P\{ n(\overline{X}_{W,n}-\overline{X}_n)/(|Q(1/n)|\vee|Q(1-1/n)|) \le x \mid F_n \}$$

converges in probability to $\mathcal{L}_\alpha(x)$. Moreover, if (3.9) holds, then the convergence is a.s.

III.4) Conclusion

This section shows that, in some non regular cases, it is possible to obtain for the generalized bootstrap results similar to the one obtained for the usual bootstrap. However, the weights for which consistency holds are quite restricted and are typically asymptotically Poisson(1). Nevertheless the methodology of resampling plans allows us sometimes to obtain necessary and sufficient conditions for the generalized bootstrap to work.

Our results in this chapter are closely related to the exploration of the undersampling scheme, which goes back to the work of Bretagnolle (1983) and has been recently investigated by Politis and Romano (1992). It should be pointed out that all the work in this section and more generally on undersampling deals only with consistency. It would be interesting to study higher order asymptotics in such non regular cases.

CHAPTER IV

PROOFS OF RESULTS OF CHAPTER I

It is convenient in all the proofs to normalize the weights $W_n := \{ W_{i,n} , 1 \le i \le n \}$ in defining

$$W_i^{(n)} := nW_{i,n} , \quad 1 \le i \le n ,$$

which ensures that $W_i^{(n)}$ is of the order of $\sigma_{W,n}$ in probability. We also abridge $S(P)$ in S.

1) Proof of Proposition I.2.1

The proof follows by induction on m. For $m = 1$, assume that $T^{(1)}(x,P)/H(x)$ is unbounded. Then choose $x_{1,n}$ such that

$$|T^{(1)}(x_{1,n},P)| \;/\; H(x_{1,n}) > n .$$

Since (I.2.11) holds, $t_{1,n} := 1 / nH(x_{1,n})$ converges to 0 as $n \to \infty$, and for n large enough,

$$Q_n := (1-t_{1,n})P + t_{1,n}\delta_{x_{1,n}}$$

is a probability measure. Moreover,

$$d_{\mathcal{H}}(Q_n,P) = t_{1,n} \sup_{h\in\mathcal{H}} |h(x_{1,n}) - E_P h(X)|$$

$$\le t_{1,n} \; (H(x_{1,n}) + E_P H(X))$$

converges to 0 as $n \to \infty$. Differentiability of T at P implies

$$T(Q_n) - T(P) = t_{1,n} DT_P(\delta_{x_{1,n}} - P) + r_n \quad \text{with}$$

$$\lim_{n\to\infty} |r_n| = 0 .$$

But this contradicts continuity of T at P implied by Fréchet-differentiability, since

$$|t_{1,n}\, DT_P(\delta_{x_{1,n}}-P)| = |t_{1,n}T^{(1)}(x_{1,n},P)|$$

$$= |T^{(1)}(x_{1,n},P)|/(nH(x_{1,n})) \geq 1 .$$

Thus, (I.2.12) holds with $p = 1$.

Assume now that (I.2.12) holds for any $p < m$ and does not hold for $p = m$. Then, choose a sequence of tuples $(x_{1,n}, \ldots, x_{m,n})_{n\geq 1}$ such that

$$T^{(m)}(x_{1,n}, \ldots, x_{m,n},P) \,/\, H(x_{1,n}) \ldots H(x_{m,n}) \geq n^m .$$

Let $t_{k,n} := 1/nH(x_{k,n})$ for $k = 1, \ldots, m$. Due to (I.2.11), each sequence $(t_{k,n})_{n\geq 1}$ converges to 0 as $n\to\infty$, and so does the sequence

$$s_n := t_{1,n} + \ldots + t_{m,n} .$$

Thus, for any large n ,

$$Q_n := (1-s_n)P + s_n V_n \quad , \text{ with}$$

$$V_n := s_n^{-1}(t_{1,n}\delta_{x_{1,n}} + \ldots + t_{m,n}\delta_{x_{m,n}})$$

is a probability measure which satisfies

$$\lim_{n\to\infty} d_{\mathcal{H}}(Q_n,P) = 0 .$$

Use the differentiability of T and notice that for $p < m$

$$D^pT_P(Q_n-P)^p = D^pT_P(t_{1,n}\delta_{x_{1,n}} + \ldots + t_{m,n}\delta_{x_{m,n}})^p$$

tends to 0 as $n \to \infty$, since (I.2.12) holds for any $p < m$. But $D^mT_P(Q_n-P)^m$ does not converge to 0 contradicting continuity of T at P , and this ends the proof □

2) Proof of Proposition I.2.2

To prove Proposition I.2.2 we obtain an upper bound for $P\big(d_{\mathcal{H}}(P_{W,n},P) > \varepsilon\big)$ which gives an upper bound for $E_{P,W} d_{\mathcal{H}}(P_{W,n},P)^q$ by an integration with respect to ε. In this proof, Pg denotes the expectation $\int_B g(x)\, dP(x)$ and $N(\varepsilon,Q)$ denotes $N_1(\varepsilon,Q,\mathcal{H})$.

We first use assumption (I.1.3) to condition on the weights W_n and to obtain independent r.v.'s, so that we can apply Pollard's (1984) type arguments.

We have, under (I.2.10) and (I.2.15),

$$\mathrm{Var}(P_{W,n}h - Ph \mid W_n) \le n^{-2} \Sigma_{i=1}^n W_i^{(n)2}\, PH^2$$

where the variance is taken under the probability P of the X_i's. Markov's inequality implies

$$\mathbb{P}\{\, |P_{W,n}h - Ph| > \varepsilon/4 \mid W_n\} \le 16\, \varepsilon^{-2}\, n^{-2}\, \Sigma_{i=1}^n W_i^{(n)2}\, PH^2 .$$

Consequently, on the event

$$A_n(\varepsilon) := \{\, a_n < \varepsilon^2 \,\} \quad \text{where}$$

$$a_n := 32\, n^{-2}\, \Sigma_{i=1}^n W_i^{(n)2}\, PH^2 ,$$

we have

$$\mathbb{P}\{\, |P_{W,n}h - Ph| > \varepsilon/4 \mid W_n \,\} \le 1/2 .$$

We can apply the symmetrization Lemma and the second symmetrization step by Pollard (1984, pp.14-15). Therefore, on $A_n(\varepsilon)$,

(2.1) $$\mathbb{P}\{\, d_{\mathcal{H}}(P_{W,n},P) > \varepsilon \mid W_n \,\} \le 4\, \mathbb{P}\{\, \sup_{h\in\mathcal{H}} |P^{\circ}_{W,n}h| > \varepsilon/4 \mid W_n \,\}$$

where $P^{\circ}_{W,n}$ denotes the signed measure

$$P^{\circ}_{W,n} := n^{-1} \Sigma_{i=1}^n \sigma_i W_i^{(n)} \delta_{X_i}$$

with $(\sigma_i)_{i\ge 1}$ an i.i.d. sequence with $P\{\sigma_i = 1\} = P\{\sigma_i = -1\} = 1/2$.

Then, we condition on the X_i's in the right hand side of (2.1) and use the covering number to obtain (see e.g. Pollard (1984) p.26 for the details)

$$\mathbb{P}\{\ \sup_{h\in\mathcal{H}}|P^{\circ}_{W,n}h| > \varepsilon/4 \mid W_n, X_1, \ldots, X_n\ \}$$

$$\le 1 \wedge N(\varepsilon/8, P_{W,n}) \max_j \mathbb{P}\{\ |P^{\circ}_{W,n}h_j| > \varepsilon/8 \mid W_n, X_1, \ldots, X_n\ \}, \tag{2.2}$$

where the functions h_j form an $\varepsilon/8$-coverage of $\mathcal{H}$ for the L_1-norm w.r.t. $P_{W,n}$. Apply Hoeffding's (1963) inequality and use the fact that H is an envelope to upper-bound (2.2) by

$$2\ N(\varepsilon/8, P_{W,n}) \exp(-\varepsilon^2 b_n) \quad \text{with} \tag{2.3}$$

$$b_n := n^2 / \{\ 128\ \Sigma_{i=1}^n W_i^{(n)2}\ H(X_i)^2\ \}. \tag{2.4}$$

From (I.2.14), we deduce that (2.3) is upper-bounded by

$$\pi_n(\varepsilon) := (\ 2A\ (P_{W,n}H/\varepsilon\ )^D \vee 2) \exp(-\varepsilon^2 b_n)\ . \tag{2.5}$$

Combine (2.1), (2.2) and (2.5) to get

$$\mathbb{P}\{\ d_{\mathcal{H}}(P_{W,n},P) > \varepsilon\ \} = E_W\ \mathbb{P}\{\ d_{\mathcal{H}}(P_{W,n},P) > \varepsilon \mid W_n\ \}$$

$$\le \mathbb{P}\{A_n(\varepsilon)^c\} + E_{W,P}\ \mathbb{I}_{A_n(\varepsilon)}(\pi_n(\varepsilon) \wedge 1)\ . \tag{2.6}$$

The bound (2.6) gives the result as follows. Observe that

$$E_{W,P}\ d_{\mathcal{H}}(P_{W,n},P)^q = q\int_0^\infty \varepsilon^{q-1}\mathbb{P}\{\ d_{\mathcal{H}}(P_{W,n},P) > \varepsilon\ \}\ d\varepsilon$$

is less than

$$q\int_0^\infty \varepsilon^{q-1}\mathbb{P}\{\ A_n(\varepsilon)^c\}\ d\varepsilon + q\int_0^\infty \varepsilon^{q-1}\ E_{W,P}\ \mathbb{I}_{A_n(\varepsilon)}(\pi_n(\varepsilon) \wedge 1)\ d\varepsilon\ . \tag{2.7}$$

The first term in (2.7) is

$$E_W\ q\int_0^\infty \varepsilon^{q-1}\ \mathbb{I}\{\ \varepsilon < (32\ n^{-2}\ \Sigma_{i=1}^n W_i^{(n)2}\ PH^2)^{1/2}\ \}\ d\varepsilon$$

$$= E_W\ (32\ n^{-2}\ \Sigma_{i=1}^n W_i^{(n)2}\ PH^2)^{q/2} = O(n^{-q/2})\ ,$$

where the last equality comes from (I.2.13) and (I.2.16).

The positivity of the r.v. $\pi_n(\varepsilon)$ ensures that the second integral in (2.7) is equal to

$$E_{W,P}\int_0^\infty \varepsilon^{q-1}\, \mathbb{1}_{A_n(\varepsilon)}\, (\pi_n(\varepsilon) \wedge 1)\, d\varepsilon$$

$$(2.8) \qquad = E_{W,P} \int_{\kappa_n^{1/2}}^\infty \varepsilon^{q-1} (\pi_n(\varepsilon) \wedge 1)\, d\varepsilon\ .$$

Cut the integral in (2.8) in $\int_{\kappa_n^{1/2}}^{\delta((\log n)/b_n)^{1/2}} + \int_{\delta((\log n)/b_n)^{1/2}}^\infty$, for some arbitrary $\delta > 0$, and use (2.5) to bound (2.8) by

$$(2.9) \qquad E_{W,P} \int_{\kappa_n^{1/2}}^{\delta((\log n)/b_n)^{1/2}} \varepsilon^{q-1}\, d\varepsilon$$

$$+ 2\, E_{W,P} \int_{\delta((\log n)/b_n)^{1/2}}^\infty \varepsilon^{q-1} \exp(-\varepsilon^2 b_n)\, d\varepsilon$$

$$+ E_{W,P} \int_{\delta((\log n)/b_n)^{1/2}}^\infty 2A\, (P_{W,n}H)^D\, \varepsilon^{q-1-D} \exp(-\varepsilon^2 b_n)\, d\varepsilon\ .$$

The first term in (2.9) is upper-bounded by

$$(2.10) \qquad E_{W,P} \int_{\kappa_n^{1/2}}^{\delta((\log n)/b_n)^{1/2}} \varepsilon^{q-1}\, d\varepsilon = (\delta^q/q)\, E_{W,P}\big((\log n)/b_n\big)^{q/2}$$

$$= O(n^{-1}\log n)^{q/2}$$

where the last equality comes from (I.2.16). To bound the last two terms in (2.9), observe that for any r, and δ large enough,

$$(2.11) \qquad E_{W,P} \int_{\delta((\log n)/b_n)^{1/2}}^\infty \varepsilon^r \exp(-\varepsilon^2 b_n)\, d\varepsilon$$

$$= (1/2)\, E_{W,P} \int_{\delta^2 \log n}^\infty y^{(r-1)/2}\, e^{-y}\, dy\, (b_n)^{-(r+1)/2}$$

$$\le (\delta^2 \log n)^{(r-1)/2} \exp(-\delta^2 \log n)\ E_{W,P}\ b_n^{-(r+1)/2}$$

so that the r.h.s. of (2.11) is

$$(\log n)^{(r-1)/2} \exp(-\delta^2 \log n)\ O(n^{-(r+1)/2}) \quad \text{as} \quad n \to \infty$$

for $r + 1 \le q$ (here, we use (I.2.15)). Choosing δ large enough ensures that the leading term in (2.9) is (2.10), and therefore, (2.4) is $O(n^{-1} \log n)^{q/2}$ which is the result if $r = q$.

If $r < q$, use the well known fact that the L_r-norm is upper-bounded by the L_q-norm in this case. □

3) Proof of Theorem I.3.1

Since (I.2.5) and (I.2.6) hold, we expand $T(P_{W,n})$ as

$$n^{1/2}(T(P_{W,n}) - T(P)) = n^{-1/2} \Sigma_{i=1}^n W_i^{(n)} T_1(i) + n^{1/2} R^{(1)}(P_{W,n}, P) \tag{3.1}$$

where we introduce the notation

$$T_1(i) := T^{(1)}(X_i, P) . \tag{3.2}$$

Similarly

$$n^{1/2}(T(P_n) - T(P)) = n^{-1/2} \Sigma_{i=1}^n T_1(i) + n^{1/2} R^{(1)}(P_n, P) . \tag{3.3}$$

Under (I.2.6), assumption (I.3.1) and Proposition I.2.2 imply

$$n^{1/2} R^{(1)}(P_{W,n}, P) = o_p(1) \quad \text{and} \quad n^{1/2} R^{(1)}(P_n, P) = o_p(1) . \tag{3.4}$$

Combining (3.1)-(3.4), we obtain (with (I.1.1))

$$n^{1/2}(T(P_{W,n}) - T(P_n)) = n^{-1/2} \Sigma_{i=1}^n W_i^{(n)} (T_1(i) - \overline{T}_{1,n}) + o_p(1)$$

with $\quad \overline{T}_{1,n} := n^{-1}\ \Sigma_{i=1}^n T_1(i)$.

Sufficient conditions for the weighted bootstrapped mean in Mason and Newton (1992) imply the result, since Proposition I.2.1 and (I.2.13) ensure

$$E_P\, T_1(i)^2 \le C_1(P)^2\, E_P H^2(X) < \infty .$$

4) Some notations and auxiliary lemmas

This subsection is devoted to the proof of some technical results that we need to prove Theorems of Chapter I. Lemma 4.1 allows us to control the remainder term in the stochastic expansions of $T(P_n)$. Lemma 4.3 gives bounds on the moments of the exchangeable weights.

As we abridged $T^{(1)}(X_i,P)$ in $T_1(i)$ in (3.2), in all the sequel, we will denote

$$T_2(i,j) := T^{(2)}(X_i,X_j,P) .$$

Similarly let $S_1(i) = S^{(1)}(X_i,P)$ be the influence function of S_n^2 .

If T is twice Fréchet-differentiable at P , (I.2.4)-(I.2.6) hold and

(4.1) $$n^{1/2}(T(P_n) - T(P)) = n^{-1/2}\Sigma_{i=1}^n T_1(i) + 2^{-1}n^{-3/2}\Sigma_{i=1}^n\Sigma_{j=1}^n T_2(i,j) + r_n$$

(4.2) $$n^{1/2}(T(P_{W,n}) - T(P)) = n^{-1/2}\Sigma_{i=1}^n W_i^{(n)} T_1(i)$$

$$+ 2^{-1}n^{-3/2}\Sigma_{i=1}^n\Sigma_{j=1}^n W_i^{(n)}W_j^{(n)}T_2(i,j) + r_{W,n}$$

where $r_n := n^{1/2}R^{(2)}(P_n,P)$ and $r_{W,n} := n^{1/2}R^{(2)}(P_{W,n},P)$.

It is convenient to introduce the notation

$$W_n(i) := nW_{i,n} - 1 = W_i^{(n)} - 1 ,$$

and to define the standardized quadratic part

(4.3) $$\sigma_{W,n}\, Q_{W,n} := n^{1/2}\Sigma_{i=1}^n W_n(i)T_1(i) +$$

$$2^{-1}n^{-3/2}\Sigma_{i=1}^n\Sigma_{j=1}^n(W_n(i)W_n(j)-W_n(i)-W_n(j))T_2(i,j) ,$$

so that (4.1) and (4.2) lead to

(4.4) $$n^{1/2}(T(P_{W,n}) - T(P_n))/\sigma_{W,n} = Q_{W,n} + R_{W,n} \quad \text{with}$$

(4.5) $$R_{W,n} := (r_n + r_{W,n})/\sigma_{W,n} \, .$$

Lemma 4.1 : **Assume that T is twice differentiable and that (I.3.4), (I.3.6) and (I.3.7) hold. Let $\rho > 0$. If assumptions of Proposition I.2.2 hold for some $q \geq \rho(2+\mu)$, then**

(4.6) $$(E_{W,P}|r_{W,n}|^{\rho})^{1/\rho} = O(n^{-(1+\lambda)/2}(\log n)^{1+\lambda/2}) \, ,$$

(4.7) $$(E_{W,P}|r_n|^{\rho})^{1/\rho} = O(n^{-(1+\lambda)/2}(\log n)^{1+\lambda/2}) \, ,$$

(4.8) $$(E_{W,P}|R_{W,n}|^{\rho})^{1/\rho} = O(n^{-(1+\lambda)/2}(\log n)^{1+\lambda/2}) \, .$$

Proof of Lemma 4.1 : (I.3.4) and (I.3.6) ensure that there exists some constants $c_{1,\rho}$ and $c_{2,\rho}$ such that

$$(E_{W,P}|r_{W,n}|^{\rho})^{1/\rho} \leq c_{1,\rho}\, n^{\rho/2}\, E_{W,P}(d_{\mathcal{H}}(P_{W,n},P)^{\rho(2+\lambda)})$$
$$+ c_{2,\rho}\, n^{\rho/2}\, E_{W,P}(d_{\mathcal{H}}(P_{W,n},P)^{\rho(2+\mu)}) \, .$$

Then apply Proposition I.2.2. to get (4.6).

(4.7) is obtained in the same way, and (4.8) is a straigthforward consequence of (4.6), (4.7) and the definition of $R_{W,n}$ in (4.5). □

Lemma 4.2 : **If Lemma 4.1 holds for some integer $\rho \geq 1$ and if the sequence of r.v.'s $(E_W|Q_{W,n}|^{\rho})_{n\geq 1}$ is tight, then**

(4.9) $$E_W(Q_{W,n} + R_{W,n})^{\rho} = E_W Q_{W,n}{}^{\rho} + o_p(n^{-1/2}) \, .$$

Proof of Lemma 4.2 : The mean value theorem ensures that for some random θ between 0 and 1 ,

(4.10) $$E_W(Q_{W,n} + R_{W,n})^{\rho} - E_W Q_{W,n}{}^{\rho} = \rho E_W R_{W,n}(Q_{W,n} + \theta\, R_{W,n})^{\rho-1} \, .$$

Use Hölder's inequality to bound the right hand side of (4.10) by

(4.11) $$\rho(E_W R_{W,n}{}^{\rho})^{1/\rho}\,(E_W(Q_{W,n}+\theta R_{W,n})^{\rho})^{(\rho-1)/\rho}\ .$$

If $\rho > 1$, we bound (4.11) by

$$\rho(E_W R_{W,n}{}^{\rho})^{1/\rho}\,2^{\rho-1}\,(E_W|Q_{W,n}|^{\rho}+E_W|R_{W,n}|^{\rho})^{1-1/\rho}$$

and apply Lemma 4.1.

If $\rho = 1$, (4.9) is a direct consequence of Lemma 4.1. □

Our next lemma shows that assumption (I.1.1) restricts the possible range for the moments of the weights.

Lemma 4.3 : **Assume that (I.1.1) holds.**

i) If $\limsup_{n\to\infty} E\,W_i^{(n)3} < \infty$ **, then**

(4.12) $$E\,W_n(i)\,W_n(j)^2 = O(n^{-1})\ .$$

ii) For any integer m **, if** $\limsup_{n\to\infty} E\,W_i^{(n)m} < \infty$ **, then for any** $i_1, i_2, \ldots, i_m$ **distinct,**

(4.13) $$E\,W_n(i_1)\ldots W_n(i_m) = O(n^{-m+1})\ .$$

Proof of Lemma 4.3 : It is an easy consequence of Proposition A.1.3 in Appendix 1. □

5) Proof of Theorem I.3.2

The proof uses the quadratic stochastic approximation (4.4) of $T(P_{W,n})$ and easy moments calculations.

Part i) : Representation (4.4) gives

(5.1) $$S_n{}^2 = E_W(Q_{W,n}+R_{W,n})^2\ .$$

Apply Lemma 4.2 to obtain

$$S_n{}^2 = E_W Q_{W,n}{}^2 + o_p(n^{-1/2})\ .$$

A careful calculation (see Appendix 3), first expanding $Q_{W,n}^2$ and then using (4.13) jointly with Remark I.2.1 and assumption (I.2.13), shows that

$$(5.2) \qquad E_W Q_{W,n}^2 = n^{-1}\Sigma_{i=1}^n T_1(i)^2 + o_p(1)$$

which is equivalent, as $n \to \infty$, to $S^2(P)$ in probability conditionally on the sample $X_1, \ldots, X_n$.

Part ii) : Proceed as in part i) to show (see Appendix 3).

$$(5.3) \qquad E_W Q_{W,n}^2 = n^{-1}\, \Sigma_{i=1}^n T_1(i)^2 + 2^{-1}\, n^{-2}\, \Sigma_{i=1}^n \Sigma_{j=1}^n T_1(i) T_2(i,j) + O_P(n^{-1}).$$

Introducing the U-statistic

$$(5.4) \qquad V_n := (2/n(n-1))\, \Sigma_{1 \le i < j \le n} \Big(2^{-1}(T_1(i)^2 + T_1(j)^2 - 2S^2(P)) + T_1(i)\, T_2(i,j) + T_1(j)\, T_2(j,i) \Big),$$

we rewrite (5.3) as

$$E_W Q_{W,n}^2 - S^2(P)$$
$$= ((n-1)/n)\, V_n + n^{-2}\, \Sigma_{i=1}^n (T_1(i)^2 + 2\, T_1(i) T_2(i,i)) + O_p(n^{-1}) .$$

Approximate V_n by its projection (see e.g. Serfling (1980)) to get the result.

Part iii) : If $S(P)^2$ is compact or Fréchet-differentiable, its influence function is

$$T^{(1)}(.,P)^2 - S^2(P) - 2 \int T^{(1)}(y,P)\, T^{(2)}(.,y,P)\, dP(y)$$

(see Pfanzagl (1985)), so that the semi-parametric efficiency bound of $S^2(P)$ is given by (I.3.10) (compare also with Beran (1984), Pfanzagl (1985, 1990)). □

6) More lemmas to prove Theorem I.3.3

The two next lemmas will be used to show that the remainder terms r_n, $r_{W,n}$ and $R_{W,n}$ in (4.1), (4.2) and (4.5) may be neglected in the Edgeworth expansion up to a $o(n^{-1/2})$.

Lemma 6.1 : **Assume that (I.2.10), (I.2.13) and (I.2.14) are satisfied, that (I.2.15) holds for some $q \geq D-1$. Then, for any sequence $(\gamma_n)_{n\geq 1}$ such that**

(6.1) $$P\{ n^{-1}\Sigma_{i=1}^{n} W_i^{(n)2} > \gamma_n \} = o(n^{-1/2}) ,$$

the bound

$$P\{ d_{\mathcal{H}}(P_{W,n},P) > \phi_n \} = o(n^{-1/2})$$

is valid for any sequence ϕ_n such that

(6.2) $$\phi_n \gg \gamma_n^{1/2}/n \vee n^{-(2q-1)/(2q+1)} \vee n^{-(2q-1)/4q} .$$

Proof of Lemma 6.1 : We first apply the bound (2.6) with $\varepsilon = \phi_n$ to obtain

$$P\{ d_{\mathcal{H}}(P_{W,n},P) > \phi_n \}$$

(6.3) $$\leq P\{ n^{-1}\Sigma_{i=1}^{n} W_i^{(n)2} > n\, \phi_n^{\;2} / 32\, PH^2 \}$$

(6.4) $$+ E_{W,P}\, \mathbb{I}\{ n^{-1}\Sigma_{i=1}^{n} W_i^{(n)2} < n\phi_n^{\;2} / 32PH^2 \}\, (\pi_n(\phi_n) \wedge 1)$$

where $\pi_n(.)$ is defined in (2.5). Condition (6.2) on ϕ_n ensures that (6.3) is $o(n^{-1/2})$. To bound (6.4), we bound $\pi_n(.)$, noticing that

(6.5) $$P_{W,n}H \leq (n^{-1}\Sigma_{i=1}^{n} W_i^{(n)2}H^2(X_i))^{1/2} = (n/128b_n)^{1/2}$$

with b_n defined in (2.4), and then, that

(6.6) $$e^{-t} \leq q^q\, e^{-q}\, t^{-q} \quad \text{for any } t \geq 0 .$$

Combining (6.5), (6.6) and expression (2.5) for $\pi_n(.)$ yields

(6.7) $$\pi_n(\phi_n) \le \Big(2\ A\ ((n/128b_n)^{1/2}/\phi_n)^D\ r^r\ e^{-r}\ (\phi_n{}^2 b_n)^{-r}\Big)$$

$$\vee\ \Big(2q^q\ e^{-q}(\phi_n{}^2 b_n)^{-q}\Big)$$

with $r := q - D/2 + 1/2$. Now, take the expectation of (6.7) and apply (I.2.15) to get

$$E\ \pi_n(\phi_n) = O(n^{-q}(\phi_n{}^{-2q-1} + \phi_n{}^{-2q}))$$

and conclude with (6.2). □

Our next lemma is an $L_{\delta+3/2}$-version of the approximation given in (I.3.8), for any $\delta > 0$. □

Lemma 6.2 : **Assume that the hypotheses of Lemma 4.1 hold for some $\rho > 3$ and (I.2.18) holds for some $q > 6$. Then**

$$E|S_n{}^2 - S^2 - n^{-1}\Sigma_{i=1}^n\ S_1(i)|^{\delta + 3/2} = O(n^{-1/2 - \delta}) .$$

Proof of Lemma 6.2 : Similarly to the proof of Theorem I.3.2, representation (4.3) (4.4) and the inequality

(6.8) $$||a|^p - |b|^p| \le p\ 2^{p-1}|a - b|^p + p\ 2^{p-1}|a - b|\ |b|^{p-1}$$

will give the result. We introduce

$$\Delta_{1,n}(r) := E_P|S_n{}^2 - E_W Q_{W,n}{}^2|^r \quad \text{and}$$

$$\Delta_{2,n} := E_P|E_W Q_{W,n}{}^2 - S^2(P) - n^{-1}\Sigma_{i=1}^n \psi(i)|^{3/2} .$$

With this notation and (6.8),

(6.9) $$\Delta_n := |\ E_P|S_n{}^2 - S^2 - n^{-1}\Sigma_{i=1}^n S_1(i)|^{\delta + 3/2}$$

$$- E_P|E_W Q_{W,n}{}^2 - S^2 - n^{-1}\Sigma_{i=1}^n S_1(i)|^{\delta + 3/2}\ |$$

$$\le (\delta + 3/2)\ 2^{\delta+1/2}\ \Delta_{1,n}(\delta+3/2)$$

$$+ (\delta + 3/2)\, 2^{\delta+1/2}\, E_P\Big(|E_W Q_{W,n}{}^2 - S^2 - n^{-1}\Sigma_{i=1}^n S_1(i)|^{\delta+1/2}$$

$$|S_n{}^2 - E_W Q_{W,n}{}^2| \Big)$$

$$\le (\delta+3/2)\, 2^{\delta+1/2}\, \Delta_{1,n}(\delta+3/2)$$

$$+ (\delta+3/2)\, 2^{\delta+1/2}\, \Delta_{2,n}{}^{(1+2\delta)/3}\, \Delta_{1,n}(3/(2-2\delta))^{(2-2\delta)/3}$$

where the last inequality comes from Hölder's inequality.

Expression (5.1) shows that

$$\text{(6.10)} \qquad \Delta_{1,n}(r) = E_P|(E_W(Q_{W,n} + R_{W,n})^2 - E_W Q_{W,n}{}^2|^r$$

$$= E_P|E_W R_{W,n}(2Q_{W,n} + R_{W,n})|^r$$

Hölder's inequality shows that (6.10) is upper-bounded by

$$E_P(\, (E_W R_{W,n}{}^2)^{r/2}\, (E_W(2Q_{W,n} + R_{W,n})^2)^{r/2}\,)$$

which, by the same argument, is smaller than

$$\Delta_{1,n}(r) \le (E_P(E_W R_{W,n}{}^2)^r)^{1/2}\, (E_P(E_W(2Q_{W,n} + R_{W,n})^2)^r)^{1/2}\, .$$

Then, for any nonnegative r.v. Z , we have $(EZ)^r \le EZ^r$, which leads to

$$\Delta_{1,n}(r) \le (E_P E_W|R_{W,n}|^{2r})^{1/2}\, (E_P\, 8^r\, |E_W Q_{W,n}{}^2 + E_W R_{W,n}{}^2|^r)^{1/2}$$

$$\le (E_{P,W}|R_{W,n}|^{2r})^{1/2}\, 16^r\, (E_P|E_W Q_{W,n}{}^2|^r + E_{P,W}|R_{W,n}|^{2r})^{1/2}\, .$$

Apply Lemma 4.1 and use assumption (I.3.23) to obtain that for $r \le \delta + 3/2$ and δ small enough,

$$\text{(6.11)} \qquad \Delta_{1,n}(r) = O(n^{-r(1+\lambda)/2}\, (\log n)^{r(2+\lambda)/2})\, .$$

For $\Delta_{2,n}$ use the trivial bound

$$\Delta_{2,n} \le 3^{3/2} \left(E_P|E_W Q_{W,n}{}^2|^{3/2} + S^3(P) + E_P|n^{-1}\Sigma_{i=1}^n \psi(X_i)|^{3/2} \right) . \tag{6.12}$$

Then, under (I.3.23), it is easy to check (see Appendix 1) that

$$E_P|E_W Q_{W,n}{}^2|^{3/2} = O(1) , \tag{6.13}$$

while Marcinkiewicz-Zygmund's Theorem with (I.3.23) and Remark I.2.1 imply

$$E_P|n^{-1}\Sigma_{i=1}^n S_1(i)|^{3/2} = O(1) \quad \text{as} \quad n\to\infty . \tag{6.14}$$

Therefore, combining the r.h.s. of (6.9), (6.11)-(6.14), we get

$$\Delta_n = O(n^{-\delta'-1/2})$$

for $\delta' > 0$ small enough, provided δ is choosen small enough. □

Our last lemma in this section shows that the studentization with S_n^2 does not change the behaviour of remainder terms in Edgeworth expansions.

Lemma 6.3 : **Assume that the hypotheses of Lemma 6.2 hold with $\rho > 4$ instead of $\rho > 3$. If $(V_n)_{n\ge1}$ is a sequence of r.v.'s such that**

$$P\{ V_n > \varepsilon_n \} = o(n^{-1/2}) \tag{6.15}$$

for some sequence $(\varepsilon_n)_{n\ge1}$ converging to 0 , then for any $m > 0$, there exists a sequence $(\eta_n)_{n\ge1}$ converging to 0 such that

$$P\{ V_n/S_n^m > \eta_n \} = o(n^{-1/2}) .$$

where S_n is given in (5.1).

Proof of Lemma 6.3 : For any $0 < \varepsilon < 1$, we have

$$P\{ V_n / S_n{}^m > \eta_n \} = P\{ (V_n/ S^m) (S/S_n)^m > \eta_n \}$$

$$\le P\{ V_n > \eta_n S^m /(1+\varepsilon) \} + P\{ S_n < S/(1+\varepsilon)^{1/m} \} .$$

Choose $\eta_n = \varepsilon_n(1+\varepsilon)/S^m$ such that the first probability in (6.16) is equal to (6.15).

From (6.11) with $r > 2/(1+\lambda)$

$$\text{(6.17)} \qquad \operatorname{Var} S_n = \operatorname{Var}\left(n^{-1} \Sigma_{i=1}^{n} S_1(i) \right) + O(n^{-1}) = O(n^{-1}) .$$

Markov's inequality yields

$$P\{ S_n - S < S((1+\varepsilon)^{-1/m}-1) \} \le P\{ |S_n - S| > S|(1+\varepsilon)^{-1/m}-1| \}$$

$$\le O(n^{-1})\, S^{-2}\, ((1-\varepsilon)^{-1/m} - 1)^{-2} = o(n^{-1/2}) . \quad \square$$

7) Proof of Theorem I.3.3

a) The main tool is the Edgeworth expansion for U-statistics of degree 2 obtained by Bickel, Götze and Van Zwet (1986) (see also Callaert, Janssen and Veraverbeke (1980)). See also Pfanzagl (1985) for the expansion of $n^{1/2}(T(P_n)-T(P))/S(P)$ in the case of a compact differentiable statistics w.r.t. the Kolmogorov distance and stronger hypotheses on the remainder term. Withers (1983) gives formal higher order expansions.

First, in formula (4.1) we show that the remainder r_n may be neglected in the Edgeworth expansion. To see this, observe that (I.2.6) and (I.3.1) with $m = 1$ lead to

$$|r_n| \le n^{1/2}\, d_{\mathcal{H}}(P_n,P)^2 \,/\, (-\log d_{\mathcal{H}}(P_n,P))^{1+\delta}$$

if $d_{\mathcal{H}}(P_n,P) \le c$ where c is some positive constant that can be chosen in $(0,1/2)$. Hence

$$\text{(7.1)} \qquad P\{ n^{1/2}|r_n| > 1/\log\log n \}$$

$$\le P\{ d_{\mathcal{H}}(P_n,P)^2 \,/\, (-\log d_{\mathcal{H}}(P_n,P))^{1+\delta} > 1 /(n \log\log n) \}$$

$$+ P\{ d_{\mathcal{H}}(P_n,P) > c \} .$$

Since $z_n^2/(\log(1/z_n))^{1+\delta} = 1/(n \log\log n)$ implies

$$z_n \sim (\log n)^{(1+\delta)/2}\, n^{-1/2}\, (\log\log n)^{-1/2} \quad \text{as} \quad n\to\infty ,$$

for n large enough, (7.1) is bounded by

$$(7.2) \qquad P\{\, n^{1/2}\, d_{\mathcal{H}}(P_n,P) > (\log n)^{(1+\delta)/2} \,\} + P\{\, d_{\mathcal{H}}(P_n,P) > c \,\} .$$

Then, under (I.3.19), one can show (follow the proof of Proposition (II.2.1)) that

$$(7.3) \qquad P\{\, n^{1/2} r_n > 1/(\log\log n) \,\} = o(n^{-1/2}) \quad \text{as } n\to\infty .$$

Combine (7.3), (4.1) and Chibishov's (1972, p.629) argument to show that the two terms Edgeworth expansion for $n^{1/2}(T(P_n) - T(P))$ is that of

$$Q_n := n^{-1/2}\Sigma_{i=1}^n T_1(i) + 2^{-1} n^{-3/2}\Sigma_{i=1}^n \Sigma_{j=1}^n T_2(i,j) .$$

Now, rewrite Q_n as a U-statistic and some remainder term

$$(7.4) \qquad Q_n \,/\, S = U_{1,n} + n^{-1/2} K_{1,1}(P)/2 + \rho_{1,n} \quad \text{with}$$

$$U_{1,n} := S^{-1}\, 2^{-1}\, n^{-3/2}\, \Sigma_{i=1}^n \Sigma_{\substack{j=1\\ j\neq i}}^n T_1(i) + T_1(j) + T_2(i,j) , \quad \text{and}$$

$$\rho_{1,n} := S^{-1}\, 2^{-1}\, n^{-3/2}\, \Sigma_{i=1}^n (2T_1(i) + T_2(i,i) - E_P T_2(i,i)) .$$

Under (I.3.16) and (I.3.19) the Edgeworth expansion for $U_{1,n}$ is given in Theorem 1.2 in Bickel, Götze and Van Zwet (1986) who show (see also Callaert, Janssen and Veraverbeke (1980)) that

$$(7.5) \qquad P\{\, U_{1,n} \le x \,\} = \Phi(x) - n^{-1/2}(K_{1,3}(P)/6)(x^2-1)\phi(x) + o(n^{-1/2})$$

uniformly in x . Then Marcinkiewicz-Zygmund's Theorem (see, for instance, Chow and Teicher (1988) Theorem 10.3.3) with Remark I.2.1 and (I.3.19) imply

$$E_P|\rho_{1,n}|^{2+\delta} = O(n^{-(2+\delta)/2}) ,$$

which implies, by Markov's inequality,

$$P\{\, n^{1/2}\rho_{1,n} > 1/\log n \,\} = o(n^{-1/2}) \,.$$

Therefore, $\rho_{1,n}$ may be neglected in the two terms Edgeworth expansion of Q_n/S, and the result follows easily from (7.5) and (7.4).

b) Denote the studentized and normalized statistics

(7.6) $$T_{n,S} := n^{1/2}(T(P_n) - T(P))/S_n \quad \text{and}$$

(7.7) $$\mathcal{S}_n := n^{1/2}(S_n^{\,2} - S^2) \,/\, S^2 \,.$$

From (4.1), our assumptions on the remainder r_n, and Theorem I.3.2.i, we deduce

$$T_{n,S} \xrightarrow{d} \mathcal{N}(0,1) \quad \text{as} \quad n \to \infty \,,$$

and (I.3.8)-(I.3.9) imply

$$\mathcal{S}_n \xrightarrow{d} \mathcal{N}(0,\; I(P)/S^4) \,.$$

To prove (I.3.25) we need to approximate $T_{n,S}$ by U-statistics for which Edgeworth expansions are known to hold. The main difficulty is to show that this approximation is legitimate and that all the remainder terms cancel asymptotically at the proper rate. We proceed in five successive approximations of $T_{n,S}$ plus the final step.

<u>First approximation</u> : from (4.1), and similarly to the proof of part i) of this theorem, we have

$$T_{n,S} = (Q_n + r_n)/S_n$$

(7.8) $$= S^{-1}\, Q_n \,(\, 1 + (S_n^{\,2} - S^2)/\, S^2)^{-1/2} + r_n \,/\, S_n \,.$$

Lemma 6.3, (7.3) and assumption (I.3.22) imply

$$P\{\, n^{1/2}\, r_n/S_n > \varepsilon_n \,\} = o(n^{-1/2})$$

for some sequence $(\varepsilon_n)_{n\geq 1}$ (maybe different from that particular one in

(7.3)) converging to 0. Therefore the two terms Edgeworth expansion for $T_{n,S}$ is that of the first term of the sum in (7.8).

Second approximation : Notice that there exists $0 < \delta < 1$ such that for any $x \in [-\delta, \delta]$,

$$|(1+x)^{-1/2} - 1 + x/2| \le x^2 .$$

Hence, we rewrite the first of the two terms in (7.8) as

$$(7.9) \qquad (Q_n/S)\,(\,1 - n^{-1/2}\mathscr{S}_n/2\,) + Q_n\,\rho_{2,n}/S$$

with, for any sequence $(\varepsilon_n)_{n\ge1}$ tending to 0 and n large enough,

$$(7.10) \qquad P\{\,n^{1/2}|Q_n\,\rho_{2,n}| > \varepsilon_n\,\}$$

$$(7.11) \qquad \le P\{\,n^{-1/2}|Q_n|\,\mathscr{S}_n^{\,2} > \varepsilon_n\,\} + P\{\,n^{-1/2}|\mathscr{S}_n| > \delta\,\}\,.$$

Using (6.17) the second term in (7.11) is $o(n^{-1/2})$.

For the first term in (7.11), first notice that

$$(7.12) \qquad P\{\,n^{-1/2}|Q_n|\,\mathscr{S}_n^{\,2} > \varepsilon_n\,\} = P\{\,|\mathscr{S}_n| > n^{1/2}\varepsilon_n\,|Q_n|^{-1/2}\,\}$$

$$(7.13) \qquad \le P\{\,|\mathscr{S}_n - \tilde{\mathscr{S}}_n| > n^{1/2}\varepsilon_n\,|Q_n|^{-1/2}/2\,\}$$

$$+ P\{|\tilde{\mathscr{S}}_n| > n^{1/2}\varepsilon_n\,|Q_n|^{-1/2}/2\,\}\,,$$

where we define

$$(7.14) \qquad \tilde{\mathscr{S}}_n := n^{-1/2}\,\Sigma_{i=1}^{n}\,S_1(i)/S^2\,.$$

Apply Markov's inequality to upper-bound the first probability in (7.13) by

$$(7.15) \qquad E_P(\,|\mathscr{S}_n - \tilde{\mathscr{S}}_n|^{3/5}\,|Q_n|^{3/10}\,)\,n^{-3/10}\,\varepsilon_n^{-3/10}\,2^{3/5}\,.$$

Then Hölder's inequality shows that (7.15) is smaller than

$$(E_P|\mathscr{S}_n - \tilde{\mathscr{S}}_n|^{3/2})^{2/5}\,(E_P|Q_n|^{1/2})^{3/5}\,n^{-3/10}\,\varepsilon_n^{-3/5}\,2^{3/5}$$

(7.16) $$= O(n^{-2(\delta+1/2)3/5(3+2\delta)})\,(E_P Q_n^2)^{3/20}\, n^{-3/5}\, \varepsilon_n^{-3/5}$$

where the last inequality comes from Lemma 6.2 and Liapunov's inequality. From (I.3.23) and Proposition I.2.1 we deduce that $E_P Q_n^2 = O(1)$ as $n\to\infty$, so that (7.16) is $O(n^{-(1/2)-8\delta/(30+20\delta)})\,\varepsilon_n^{-3/5}$.

For the second probability in (7.13), Markov's inequality and Hölder's inequality give the bound

$$E_P(|\tilde{\mathscr{S}}_n|^{3/2}|Q_n|^{3/4})\, n^{-3/4}\, \varepsilon_n^{-3/2}\, 2^{3/2}$$

(7.17) $$\le (E_P|\tilde{\mathscr{S}}_n|^2)^{4/3}\,(E_P|Q_n|^3)^{1/4}\, n^{-3/4}\, \varepsilon_n^{-3/2}\, 2^{3/2} .$$

Under (I.3.23) we have $E_P\tilde{\mathscr{S}}_n^2 = O(1)$ as $n\to\infty$. Since Q_n is a V-statistic, Lemmas 6.1 and 5.2.2.A in Serfling (1980) show that $E_P|Q_n|^3 = O(1)$ as $n\to\infty$, so that (7.17) is $O(n^{-3/4}\varepsilon_n^{-3/2})$.

Therefore, (7.12) is $o(n^{-1/2})$ (as $n\to\infty$) for, says, $\varepsilon_n = 1/\log n$ and so is (7.10). Consequently, the Edgeworth expansion for (7.9) and (7.6) is that of $Q_n\,(1 - n^{-1/2}\mathscr{S}_n/2)/S$.

Third approximation : Definition (7.7) and the expansion (I.3.8) suggest to approximate $\mathscr{S}_n$ by $\tilde{\mathscr{S}}_n$ defined in (7.14). To prove that this approximation is valid in the Edgeworth expansion, first notice that for any $(\varepsilon_n)_{n\ge1}$ and $\delta > 0$,

(7.18) $$P\{\, n^{1/2}|Q_n|\, n^{-1/2}\, |\mathscr{S}_n - \tilde{\mathscr{S}}_n| > \varepsilon_n \}$$

$$\le E_P|Q_n(\mathscr{S}_n - \tilde{\mathscr{S}}_n)|^{1+\delta}\, \varepsilon_n^{-1-\delta}$$

$$\le (E_P|Q_n|^{3(1+\delta)})^{1/3}\,(E_P|\mathscr{S}_n - \tilde{\mathscr{S}}_n|^{3(1+\delta)/2})^{2/3}\, \varepsilon_n^{-1-\delta} .$$

Since (I.3.22) holds, Lemma 6.2 shows that for δ small enough

$$E_P|\mathscr{S}_n - \tilde{\mathscr{S}}_n|^{3(1+\delta)/2} = o(n^{-(1/2)-\delta}) .$$

Next, Lemma 5.2.2.A in Serfling (1980) implies

$$E_P|Q_n|^{3(1+\delta)} = O(1) \quad \text{as} \quad n\to\infty ,$$

so that (7.18) is upper-bounded by $\varepsilon_n^{-1-\delta}o(n^{-(1/2)-\delta})$, and there exists a sequence $(\varepsilon_n)_{n\geq 1}$ converging to 0 for which (7.18) is $o(n^{-1/2})$. Then the same argument as in Chibishov (1972, p.629), shows that the two terms Edgeworth expansion for $Q_n(1 - n^{-1/2}\mathcal{G}_n/2)/S$ (and so, that of (7.6)) is that of

$$(7.19) \qquad Q_n(1 - n^{-1/2}\tilde{\mathcal{G}}_n/2)/S$$

$$(7.20) \qquad = \Big(n^{-1/2} \Sigma_{i=1}^n T_1(i) + 2^{-1}n^{-3/2} \Sigma_{i=1}^n\Sigma_{j=1}^n T_2(i,j) \Big)$$

$$(1 - 2^{-1}\, n^{-1}\, S^{-2}\, \Sigma_{i=1}^n\, S_1(i)) \,/\, S.$$

<u>Fourth approximation</u> : The next step is to approximate (7.20) by a U-statistic of degree two. For this, we introduce the notations

$$V_{1,n} := n^{-2}\, \Sigma_{i=1}^n\Sigma_{j=1}^n T_2(i,j) ,$$

$$U_{1,n} := (2/n(n-1))\, \Sigma_{1\leq i<j\leq n} T_2(i,j) \quad \text{and}$$

$$\overline{S_{1,n}} := n^{-1}\, \Sigma_{i=1}^n\, S_1(i) .$$

Clearly $U_{1,n}$ is a U-statistic which is close to $V_{1,n}$. Observe that

$$V_{1,n} - ((n-1)/n)U_{1,n} = n^{-2}\, \Sigma_{i=1}^n\, T_2(i,i) .$$

Consequently, Markov's inequality and Marcinkiewicz-Zygmund's Theorem imply for any sequence $(\varepsilon_n)_{n\geq 1}$ and any $\rho > 0$,

$$(7.21) \qquad P\{\, n|V_{1,n} - ((n-1)/n)U_{1,n}|\; |\overline{S_{1,n}}| > \varepsilon_n \,\}$$

$$\leq \varepsilon_n^{-\rho}\, n^\rho\, E_P|V_{1,n} - ((n-1)/n)U_{1,n}|^\rho |\overline{S_{1,n}}|^\rho$$

$$\leq \varepsilon_n^{-\rho}\, n^\rho\, (E_P|V_{1,n} - ((n-1)/n)U_{1,n}|^{2\rho})^{1/2}(E_P|\overline{S_{1,n}}|^{2\rho})^{1/2}$$

$$= \varepsilon_n^{-\rho}\, n^{-\rho/2}\, O(1) \quad \text{as} \quad n\to\infty .$$

Under (I.3.23), we take $5/4 > \rho > 1$ to ensure the existence of a sequence $(\varepsilon_n)_{n\geq 1}$ converging to 0 such that (7.21) is $o(n^{-1/2})$ as $n\to\infty$. Then, similarly,

$$(7.22) \qquad P(n|\overline{S_{1,n}}||U_{1,n}| > \varepsilon_n) \leq \varepsilon_n^{-1-\delta} n^{1+\delta} E|\overline{S_{1,n}} U_{1,n}|^{1+\delta}$$

$$(7.23) \qquad \leq \varepsilon_n^{-1-\delta} n^{1+\delta} (E_P|\overline{S_{1,n}}|^{2(1+\delta)/(1-\delta)})^{(1-\delta)/2} (E_P|U_{1,n}|^2)^{(1+\delta)/2} ,$$

and Lemma 5.2.2.B in Serfling (1980) shows that $E_P|U_{1,n}|^2 = O(n^{-2})$.

Thus (7.23) is

$$(7.24) \qquad \varepsilon_n^{-1-\delta} n^{1+\delta} n^{-(1+\delta)/2} n^{-(1+\delta)} O(1)$$

and there exists a sequence $(\varepsilon_n)_{n\geq 1}$ converging to 0 such that (7.24), and therefore (7.22), are $o(n^{-1/2})$. This implies that there exists a sequence $(\varepsilon_n)_{n\geq 1}$ converging to 0 and such that

$$(7.25) \qquad P(n|V_{1,n} \overline{S_{1,n}}| > \varepsilon_n) = o(n^{-1/2}) .$$

We infer from (7.25) and (7.19)-(7.20) that the two terms Edgeworth expansion of (7.6) is that of

$$V_{2,n} := S^{-1}n^{-1/2} \Sigma_{i=1}^n T_1(i) + 2^{-1}S^{-1}n^{-3/2} \Sigma_{i=1}^n \Sigma_{j=1}^n T_2(i,j)$$

$$- 2^{-1}S^{-3} n^{-1} \Sigma_{i=1}^n S_1(i) n^{-1/2} \Sigma_{i=1}^n T_1(i) .$$

<u>Fifth approximation</u> : Now, rewrite $V_{2,n}$ as a U-statistic, $U_{2,n}$, plus an empirical mean, $M_{2,n}$, with

$$U_{2,n} := 2^{-1}S^{-1} n^{-3/2} \Sigma_{i=1}^n \Sigma_{\substack{j=1 \\ j\neq i}}^n (T_1(i) + T_1(j) + T_2(i,j)$$

$$- 2^{-1}S^{-2}S_1(i)T_1(j) - 2^{-1}S^{-2}S_1(j)T_1(i)) \quad \text{and}$$

$$M_{2,n} := 2^{-1}S^{-1}n^{-3/2} \Sigma_{i=1}^n 2T_1(i) + T_2(i,i) + S^{-2} S_1(i) T_1(i) .$$

Obviously there exists a sequence $(\varepsilon_n)_{n\geq 1}$ converging to 0 such that

$$P\{ n^{1/2}|M_{2,n} - \nu_{2,n}| > \varepsilon_n \} = o(n^{-1/2}) \quad \text{where}$$

$$\nu_{2,n} := E_P\{T_2(i,i) - S^{-2} S_1(i) T_1(i)\} / (n^{1/2} 2S)$$

$$= n^{-1/2} K'_{1,1}(P)/2 ,$$

so that the Edgeworth expansion of (7.6) is that of $U_{2,n} + n^{-1/2} K'_{1,1}(P)/2$.

Final Step : Apply Theorem 1.2 in Bickel, Götze and Van Zwet (1986) to prove that with

$$K'_{1,3}(P) = S^{-3}(E_P T^{(1)}(X,P)^3$$
$$+ 3 E_P\{ T^{(1)}(X_1,P)T^{(1)}(X_2,P)T^{(2)}(X_1,X_2,P)\}$$
$$- (3/2S^2) E_P\{T^{(1)}(X_1,P)T^{(1)}(X_2,P)^2 S^{(1)}(X_1,P) \}$$
$$- (3/2S^2) E_P\{T^{(1)}(X_1,P)^2 T^{(1)}(X_2,P)^2 S^{(1)}(X_2,P) \})$$

(in order to get the expression of $K'_{1,3}(P)$ given in (I.3.14), just replace $S^{(1)}(X,P)$ by its value given in (I.3.9)) the Edgeworth expansion for $U_{2,n}$ is given by

$$\sup_{-\infty<x<\infty}|P\{ U_{2,n} \le x \} - \Phi(x) + K'_{1,3}(P)/(6n^{1/2}) \phi(x)(x^2-1)| = o(n^{-1/2}) \tag{7.26}$$

From (7.26) we easily deduce the Edgeworth expansion for $U_{2,n} + n^{-1/2} K'_{1,1}(P)/2$ (which is also that of (7.6)), which is nothing else than (I.3.25). □

8) Proof of Theorem I.3.4

i) To prove (I.3.31), use the representation (4.4) to obtain

$$\mu_{1,n}/\sigma_{W,n} = n^{1/2} n^{1/2} E_W(T(P_{W,n}) - T(P_n))/\sigma_{W,n}$$

$$= n^{1/2} E_W Q_{W,n} + n^{1/2} E_W R_{W,n} . \tag{8.1}$$

We shall show that the term $n^{1/2}E_W R_{W,n}$ cancels asymptotically while $n^{1/2}E_W Q_{W,n}$ is approximatively equal to $E_P T_2(i,i)\ E\ W_n(1)^2/2$.

We actually prove a result a little stronger than (I.3.31) but which will be useful to prove (I.3.33), namely that

(8.2) $$P\{\ n^{1/2}|E_W R_{W,n}| > \varepsilon_n\ \} = o(n^{-1/2})$$

for some sequence $(\varepsilon_n)_{n\geq 1}$ converging to 0 . To see this, bound the probability in (8.2) by

$$\varepsilon_n^{-\rho} n^{\rho/2} E_P |E_W R_{W,n}|^{\rho}$$

(8.3) $$\leq \varepsilon_n^{-\rho} n^{\rho/2} E_{P,W} |R_{W,n}|^{\rho}$$

for any $\rho \geq 1$. Then apply Lemma 4.1 to bound (8.3) by

$$\varepsilon_n^{-\rho}\ n^{-\rho\lambda/2}\ (\log\ n)^{\rho(2+\lambda)/2}$$

and conclude in choosing ρ such that $\rho\lambda > 1$ with (I.3.30). Then, Lemma 4.3.i and (4.3) imply

$$n^{1/2}E_W Q_{W,n} = (1/2)\ n^{-1}\ \Sigma_{i=1}^{n} T_2(i,i) + O(n^{-2})\ \Sigma_{1\leq i<j\leq n} T_2(i,j)$$

where the $O(n^{-2})$-term is deterministic. Then, obviously, there exists a sequence $(\varepsilon_n)_{n\geq 1}$ converging to 0 with

(8.4) $$P\{\ |n^{-1}\Sigma_{i=1}^{n} T_2(i,i) - E_P T_2(i,i)| > \varepsilon_n\ \} = o(n^{-1/2})$$

(this is a straightforward application of Markov's inequality and Marcinkiewicz-Zygmund's Theorem) and

(8.5) $$P\{\ |n^{-2} \sum_{1\leq i<j\leq n} T_2(i,j)| > \varepsilon_n\ \} \leq \varepsilon_n^{-2}\ E(n^{-2} \sum_{1\leq i<j\leq n} T_2(i,j))^2$$

(8.6) $$\leq \varepsilon_n^{-1}\ O(n^{-1})$$

by Lemma 5.2.1.A in Serfling (1980). Now, combine (8.2), (8.4) and (8.5)-(8.6) to infer

$$\mu_{1,n} = E_P T_2(i,i)\ E\ W_n(1)^2/2 + \rho_{1,n}$$

where $\rho_{1,n}$ is a remainder satisfying

(8.7) $$P\{\ |\rho_{1,n}| > \varepsilon_n\ \} = o(n^{-1/2})$$

for some sequence $(\varepsilon_n)_{n\geq 1}$ converging to 0 .

ii) We first study $\mu_{3,n}$: for any sequence $(\varepsilon_n)_{n\geq 1}$ we have for any $\rho > 0$

(8.8) $$P\{\ |\mu_{3,n} - n^{1/2}\sigma_{W,n}^3\ E_W Q_{W,n}^3| > \varepsilon_n\ \}$$

(8.9) $$\leq \varepsilon_n^{-\rho}\ E_P|\mu_{3,n} - n^{1/2}\sigma_{W,n}^3\ E_W Q_{W,n}^3|^\rho .$$

Use representation (4.4) to rewrite (8.9) divided by $\sigma_{W,n}^3$ as

$$\varepsilon_n^{-\rho}\ n^{\rho/2}\ E_P|E_W R_{W,n}(Q_{W,n}^2 + Q_{W,n}R_{W,n} + R_{W,n}^2)|^\rho$$

$$\leq \varepsilon_n^{-\rho}\ n^{\rho/2}\ E_{P,W}|R_{W,n}Q_{W,n}^2 + Q_{W,n}R_{W,n}^2 + R_{W,n}^3|^\rho$$

(8.10) $$\leq 3\ 2^\rho\ \varepsilon_n^{-\rho}\ n^{\rho/2}\ (\ E_{P,W}|R_{W,n}Q_{W,n}^2|^\rho + E_{P,W}|R_{W,n}^2 Q_{W,n}|^\rho + E_{P,W}|R_{W,n}|^{3\rho}\ ) .$$

Then apply Hölder's inequality, Lemma 4.1 and Appendix 2 to obtain

$$E_{P,W}|R_{W,n}Q_{W,n}^2|^\rho \leq (E_{P,W}|R_{W,n}|^{3\rho})^{1/3}\ (E_{P,W}|Q_{W,n}|^{3\rho})^{2/3}$$

$$= O(n^{-(1+\lambda)\rho/2}\ (\log\ n)^{\rho(1+\lambda/2)}) = o(n^{-(\rho+1)/2}).$$

if $\rho = [3/\lambda]/3$. By the same arguments we have

(8.11) $$E_{P,W}|R_{W,n}^2 Q_{W,n}|^\rho \leq (E_{P,W}|R_{W,n}|^{3\rho})^{2/3}\ (E_{P,W}|Q_{W,n}|^{3\rho})^{1/3} .$$

Then, (I.3.30) ensures that we can choose $\rho = [3/\lambda]/3$ so that Lemma 4.1 and Appendix 2 yield that (8.11) is $O(n^{-(1+\lambda)\rho}\ (\log\ n)^{\rho(1+\lambda/2)})$. Hence, (8.10) is $O(n^{-\lambda\rho/2}\ (\log\ n)^{\rho(1+\lambda/2)})\ \varepsilon_n^{-\rho} = o(n^{-1/2})$. Thus, there exists a sequence $(\varepsilon_n)_{n\geq 1}$ such that (8.8) is $o(n^{-1/2})$ and ε_n converges to 0 as $n\to\infty$.

Then, a careful, but easy, calculation using Lemmas 4.2 and 4.3 shows that

$$n^{1/2}\,\sigma_{W,n}^{3}\,E_W Q_{W,n}^{3} = \bar{\mu}_{3,n} + \rho_{3,n} \quad \text{with}$$

$$\bar{\mu}_{3,n} := E_W W_n(1)^3\, E_P T_1(i)^3 + 3E_W(W_n(1)^2 W_n(2)^2)\, E_P T_1(1) T_1(2) T_2(1,2) \tag{8.12}$$

$$+ (3/2)\, E_W W_n(1)^2 W_n(2)^2\, E_P T_1(1)^2\, E_P T_2(1,1)$$

and where $\rho_{3,n}$ is such that

$$E_{P,W}\,\rho_{3,n}^{2+\delta} = O(n^{-1-\delta/2}) \tag{8.13}$$

for some $\delta > 0$.

Obviously (8.12), (8.13) and the fact that (8.8) is $o(n^{-1/2})$ implies that for some sequence $(\varepsilon_n)_{n\geq 1}$ converging to 0 (possibly different from that in (8.8)), one has

$$P\{\, |\mu_{3,n} - \bar{\mu}_{3,n}| > \varepsilon_n \,\} = o(n^{-1/2}) \, . \tag{8.14}$$

Now, observe that

$$\bar{\mu}_{3,n} - (3/2) S^2 E_W W_n(1)^2 W_n(2)^2\, E_P T_2(1,1)$$

$$= E_W W_n(1)^3\, E_P T_1(1)^3 + 3E_W W_n(1)^2 W_n(2)^2\, E_P T_1(1) T_1(2) T_2(1,2) \tag{8.15}$$

which leads to (I.3.32) using (I.3.28) and Theorem 3.2 to estimate S^2.

iii) Applying Theorem 1 in Abramovitch and Singh (1985), (I.3.33) holds if, for some sequence $(\varepsilon_n)_{n\geq 1}$ converging to 0 ,

$$P\{\, |K_{1,n} - K_{1,1}(P)| > \varepsilon_n \,\} = o(n^{-1/2}) \quad \text{and} \tag{8.16}$$

$$P\{\, |K_{3,n} - K_{1,3}(P)| > \varepsilon_n \,\} = o(n^{-1/2}) \, . \tag{8.17}$$

But (8.16) follows from (8.7), and combining (8.15), (8.14), (8.13) and Lemma 6.3 yields (8.17). □

9) Proof of Theorem I.3.5

Using notations of the proof of Theorem I.3.4, we just add the index W , I or J according to the weights used. Since all the conditions of Theorem I.3.4 except (I.3.28) hold, we deduce from (8.12) that

$$\bar{\mu}_{3,n}(I) - \bar{\mu}_{3,n}(J) = E(I_n(1)^3 - J_n(1)^3)\, E_P T_1(1)^3$$

with some obvious notations. This leads to (I.3.39).

Now, (I.3.40), i.e. the fact that we can replace $K'_{1,1}(P)$ and $K'_{1,3}(P)$ by their estimators in the Edgeworth expansion given by (I.3.25), comes from all the results proved in section 8, and Theorem 1 in Abramovitch and Singh (1985) and Lemma 6.3 when we need to standardize by an estimator S_n . □

CHAPTER V

PROOFS OF RESULTS OF CHAPTER II

1) Proof of results of section II.2

We use similar techniques as in the proof of Proposition I.2.2. We first observe the obvious equality

$$\Sigma_{i=1}^{n} (W_{i,n}-1/n)\ h(X_i)$$

$$= n^{-1}\Sigma_{i=1}^{n}(Y_{i,n}-1)h(X_i) + \overline{Y}_n^{-1}(1-\overline{Y}_n)\ n^{-1}\Sigma_{i=1}^{n}Y_{i,n}h(X_i)\ .$$

Hence, if we define

$$\delta_{1,n} := \sup_{h\in\mathcal{H}}|n^{-1}\Sigma_{i=1}^{n}(Y_{i,n}-1)h(X_i)|\ ,$$

$$\delta_{2,n} := \sup_{h\in\mathcal{H}}|n^{-1}\Sigma_{i=1}^{n}Y_{i,n}h(X_i)| \quad \text{and}$$

$$\chi_n := |(1-\overline{Y}_n)/\overline{Y}_n|\ ,$$

we have

$$\text{(1.1)} \qquad E(d_{\mathcal{H}}(P_{W,n},P_n)^{\rho} \mid P_n) \le 2^{\rho}\ (\ E(\delta_{1,n}{}^{\rho}|P_n) + 2^{\rho}\ E(\chi_n{}^{\rho}\delta_{2,n}{}^{\rho}|P_n)\ .$$

To bound $E(\delta_{1,n}{}^{\rho}|P_n)$ we first bound the tail probability $P\{\delta_{1,n}\ge x \mid P_n\}$.

Since the $Y_{i,n}$ are i.i.d., if we denote $\sigma_{Y,n}{}^{2} := \text{Var}\ Y_{i,n}$,

$$\text{Var}(n^{-1}\Sigma_{i=1}^{n}(Y_{i,n}-1)h(X_i) \mid P_n) \le n^{-2}\sigma_{Y,n}{}^{2}\ \Sigma_{i=1}^{n}H(X_i)^2$$

which converges a.s. to 0 , since $E\ H(X_i)^2 < \infty$. Then Chebyshev's inequality yields, for any $h \in \mathcal{H}$,

$$P\{\ |n^{-1}\Sigma_{i=1}^{n}(Y_{i,n}-1)h(X_i)| \le x \mid P_n\ \} \ge 1/2\ ,$$

on the set

$$A(P_n) := \{ x : x^2 \geq 2\,\sigma_{Y,n}^2\, n^{-2}\, \Sigma_{i=1}^n H(X_i)^2 \} .$$

Following Pollard (1984, chapter 2), on $A(P_n)$, we have

$$P\{ \delta_{1,n} \geq x \mid P_n \} \leq 4\, P\{ \sup_{h\in\mathcal{H}} |n^{-1}\Sigma_{i=1}^n \varepsilon_i Y_{i,n} h(X_i)| \geq x/4 \mid P_n \}$$

where the ε_i are i.i.d. with distribution

$$P\{ \varepsilon_i = 1 \} = P\{ \varepsilon_i = -1 \} = 1/2 ,$$

and are independent on both the X_i's and the $Y_{i,n}$'s .

As in section II.6, let $P_{Y,n}$ be the nonnegative measure

$$P_{Y,n} = n^{-1}\, \Sigma_{i=1}^n Y_{i,n} \delta_{X_i}$$

and let $\{ h_j : 1 \leq j \leq N_1(x/8, P_{Y,n}, \mathcal{H}) \}$ be an L_1-covering of $\mathcal{H}$ such that

$$\sup_{h\in\mathcal{H}} \min_j P_{Y,n}|h-h_j| \leq x/8 .$$

Then, on $A(P_n)$, we have

$$P\{ \delta_{1,n} \geq x \mid P_n, Y_{1,n}, \ldots, Y_{n,n} \}$$

$$\leq 4\, N_1(x/8, P_{Y,n}, \mathcal{H})\, \max_j P\{ |n^{-1}\Sigma_{i=1}^n \varepsilon_i Y_{i,n} h_j(X_i)| \geq x/8 \mid P_n, Y_{1,n}, \ldots, Y_{n,n} \} . \tag{1.2}$$

Using Hoeffding's inequality we get

$$P\{ |n^{-1}\Sigma_{i=1}^n \varepsilon_i Y_{i,n} h_j(X_i)| \geq x/8 \mid P_n, Y_{1,n}, \ldots, Y_{n,n} \}$$

$$\leq 2 \exp(-x^2 n/(128 n^{-1}\Sigma_{i=1}^n Y_{i,n}^2 h_j(X_i)^2))$$

$$\leq 2 \exp(-x^2 n/(128 n^{-1}\Sigma_{i=1}^n Y_{i,n}^2 H(X_i)^2)) . \tag{1.3}$$

Then (1.2) and (1.3) give the bound

$$P\{ \delta_{1,n} \ge x \mid P_n \}$$

$$\le 8\ E(N_1(x/8, P_{Y,n}, \mathcal{H}) \exp(-x^2 n/(128 n^{-1} \Sigma_{i=1}^n Y_{i,n}^2 H(X_i)^2) \mid P_n)$$

on $A(P_n)$.

Since $\mathcal{H}$ is a polynomial class and $P_{Y,n}$ is a measure a.s. finite, and since

$$P_{Y,n}(-\infty,\infty) = n^{-1} \Sigma_{i=1}^n Y_{i,n} ,$$

we have (using (I.2.14))

$$N_1(x/8, P_{Y,n}, \mathcal{H}) = N_1(x/8\overline{Y}_n, P_{Y,n}/\overline{Y}_n, \mathcal{H}) \le A(8 P_{Y,n} H/x)^D \vee 1 .$$

Therefore, on the set $A(P_n)$, we have

$$\text{(1.4)} \qquad P\{ \delta_{1,n} \ge x \mid P_n \} \le 8\ E_P(A(8 P_{Y,n} H/x)^D \vee 1) \exp(-x^2 b_n) \mid P_n)$$

with

$$b_n := n/(128 n^{-1} \Sigma_{i=1}^n Y_{i,n}^2 H(X_i)^2) .$$

On the set $A(P_n)^c$, we use the trivial bound

$$\text{(1.5)} \qquad P\{ \delta_{1,n} \ge x \mid P_n \} \le 1 .$$

Using (1.4) and (1.5), we obtain an inequality for

$$\text{(1.6)} \qquad E(\delta_{1,n}^{\rho} \mid P_n) = \rho \int_0^\infty x^{\rho-1} P\{ \delta_{1,n} \ge x \mid P_n \}\ dx .$$

Let us introduce

$$\xi_n := (8\ \sigma_{Y,n}^2\ n^{-2}\ \Sigma_{i=1}^n H(X_i)^2)^{1/2} \quad \text{and}$$

$$a_n := 8\ n^{-1} \Sigma_{i=1}^n Y_i H(X_i) = 8\ P_{Y,n} H .$$

Then, (1.4)-(1.5) yield

(1.7) $$E(\,\delta_{1,n}{}^{\rho} \mid P_n\,)$$

$$\le \rho \int_0^{\xi_n} x^{\rho-1}\,dx + \rho \int_0^{\infty} x^{\rho-1} E(\,(1\wedge A(a_n/x)^D+1)\exp(-x^2 b_n) \mid P_n\,)\,dx\,.$$

In the second integral of the r.h.s. of (1.7), we permute the expectation and the integral (this is valid since the functions whose we take the expectation and that we integrate are nonnegative), to bound it by

(1.8) $$E(\int_0^{\eta} x^{\rho-1} dx \mid P_n\,) + E(\int_{\eta}^{\infty} x^{\rho-1}(A(a_n/x)^D+1)\exp(-x^2 b_n)\,dx \mid P_n\,)$$

for any $\eta \ge 0$, possibly random.

Observe that for any $r \ge 0$,

(1.9) $$\int_{\eta}^{\infty} x^r \exp(-x^2 b_n)\,dx = \int_{\eta^2 b_n}^{\infty} y^{(r-1)/2} e^{-y}\,dy\; b_n{}^{-(r+1)/2}\,,$$

and moreover, integrating by parts,

(1.10) $$\int_{\eta^2 b_n}^{\infty} (y^{(r-1)/2} - y^{(r-3)/2}(r-1)/2)\; e^{-y} dy = (\eta^2 b_n)^{(r-1)/2} \exp(-\eta^2 b_n).$$

We choose $\eta = (c(\log n)/b_n)^{1/2}$, where c is a constant precised hereafter. Then, (1.9) and (1.10) imply that for n large (i.e. $n \ge N_1$ where N_1 depends only on $c \log n$ and r and therefore is deterministic), we have

(1.11) $$\int_{\eta}^{\infty} x^r \exp(-x^2 b_n)\,dx \le 2(\eta^2 b_n)^{(r-1)/2}\, b_n{}^{-(r+1)/2} \exp(-\eta^2 b_n)$$

$$= 2(c \log n)^{(r-1)/2} \exp(-c \log n)\, b_n{}^{-(r+1)/2} =: i_n(r)\,.$$

Then, from (1.8) and (1.11), we deduce that (1.7) is upper-bounded by

$$\rho\, \xi_n{}^{\rho} + E(\eta^{\rho} \mid P_n) + \rho\, A\, E(\,a_n{}^D i_n(\rho-1-D) + i_n(\rho-1) \mid P_n\,)$$

(1.12) $$= \rho\, \xi_n{}^{\rho} + E(\eta^{\rho} \mid P_n)$$

$$+ 2\ \rho\ A\ E(\ a_n^{\ D}(c\ \log\ n)^{(\rho-2-D)/2}\exp(-c\log\ n)\ b_n^{\ -(\rho-D)/2}\ \mid P_n\ )$$

$$+ 2\ \rho\ A\ E(\ (c\ \log\ n)^{(\rho-2)/2}\ \exp(-c\ \log\ n)\ b_n^{\ -\rho/2}\ \mid P_n\ )\ .$$

We bound each term of (1.12) as follows :

$$E(\eta^{\rho}\ \mid P_n)$$

$$= E(\ (c\ \log\ n)^{\rho/2}\ (128/n)^{\rho/2}\ (n^{-1}\Sigma_{i=1}^{n} Y_{i,n}^{\ 2} H(X_i)^2)^{\rho/2}\ \mid P_n\ )$$

$$\le (128cn^{-1}\log\ n)^{\rho/2}\ MZ(Y_{i,n}^{\ 2\rho},1/2)\ (n^{-1}\Sigma_{i=1}^{n} H(X_i)^{2\rho/(\rho-1)})^{(\rho-1)/2}$$

(here we use Hölder's inequality, and then, Marcinkiewicz-Zygmund's inequality), where for any r.v. Z_i , $MZ(Z_i,s)$ denotes the constant in the Marcinkiewicz-Zygmund type inequality

$$\sup_{n\ge1}\ E|n^{-1}\Sigma_{i=1}^{n}\ Z_i|^{s} \le MZ(Z_i,s)$$

Then we have

$$E(\ a_n^{\ D} b_n^{\ -(\rho-D)/2}\ \mid P_n\ )$$

$$= (128/n)^{(\rho-D)/2}\ 8^{D}$$

$$E\Big(\ (n^{-1}\ \Sigma_{i=1}^{n}\ Y_{i,n}\ H(X_i))^{D}(n^{-1}\Sigma_{i=1}^{n} Y_{i,n}^{\ 2} H(X_i)^2)^{(\rho-D)/2}\ \mid P_n\ \Big)$$

$$\le (128/n)^{(\rho-D)/2}\ 8^{D}\ \Big(\ E((n^{-1}\ \Sigma_{i=1}^{n}\ Y_{i,n}\ H(X_i))^{2D}\ \mid P_n\ \Big)^{1/2}$$

and

$$\Big(E((n^{-1}\Sigma_{i=1}^{n} Y_{i,n}^{\ 2} H(X_i)^2)^{(\rho-D)}\ \mid P_n\ \Big)^{1/2}$$

$$\le (128/n)^{(\rho-D)/2}\ 8^{D}\ \Big(\ E(n^{-1}\ \Sigma_{i=1}^{n}\ Y_{i,n}^{\ 2})^{D}\ \Big)^{1/2}$$

$$\Big(n^{-1}\ \Sigma_{i=1}^{n} H(X_i)^2\Big)^{D/2}\ E(n^{-1}\ \Sigma_{i=1}^{n}\ Y_{i,n}^{\ 4})^{(\rho-D)/2}$$

$$(n^{-1}\Sigma_{i=1}^{n} H(X_i)^4)^{(\rho-D)/4}$$

$$\leq (128/n)^{(\rho-D)/2}\, 8^D\, MZ(Y_{i,n}^2, D)^{1/2}\, MZ(Y_{i,n}^4, (\rho-D)/4)$$

$$(n^{-1}\, \Sigma_{i=1}^n H(X_i)^2)^{D/2} (n^{-1}\Sigma_{i=1}^n H(X_i)^4)^{(\rho-D)/4} \ ,$$

and similarly,

$$E(\, b_n^{-\rho/2} \mid P_n \,) = (128/n)^{\rho/2}\, E((n^{-1}\Sigma_{i=1}^n Y_{i,n}^2 H(X_i)^2)^{\rho/2} \mid P_n)$$

$$\leq (128/n)^{\rho/2}\, E((n^{-1}\Sigma_{i=1}^n Y_{i,n}{}^4)^{\rho/4}\, (n^{-1}\, \Sigma_{i=1}^n H(X_i)^4)^{\rho/4}$$

$$\leq (128/n)^{\rho/2}\, MZ(Y_{i,n}^{\ 4}, \rho/4)\, (n^{-1}\, \Sigma_{i=1}^n H(X_i)^4)^{\rho/4} \ .$$

Therefore, for some constants k_1, k_2, k_3, k_4,

$$E(\, \delta_{1,n}^{\ \rho} \mid P_n \,) \leq k_1\, (n^{-2}\Sigma_{i=1}^n H(X_i)^2)^{\rho/2} \tag{1.13}$$

$$+ k_2\, (n^{-1} \log n)^{\rho/2}\, (n^{-1}\Sigma_{i=1}^n H(X_i)^{2\rho/(\rho-1)})^{(\rho-1)/2}$$

$$+ k_3\, (\log n)^{(\rho-2-D)/2}\, n^{-c}\, n^{-(\rho-D)/2}\, (n^{-1}\Sigma_{i=1}^n H(X_i)^2)^{D/2}$$

$$(n^{-1}\Sigma_{i=1}^n H(X_i)^4)^{(\rho-D)/2}$$

$$+ k_4\, (\log n)^{(\rho-2)/2}\, n^{-c}\, n^{-\rho/2}\, (n^{-1}\Sigma_{i=1}^n H(X_i)^4)^{\rho/4} \ .$$

It remains us to bound $E(\chi_n^{\ \rho}\delta_{2,n}^{\ \rho} \mid P_n)$. For this, note that

$$|\chi_n \delta_{2,n}| \leq |(\, 1 - \bar{Y}_n)\, (n^{-1}\Sigma_{i=1}^n Y_{i,n}^2)^{1/2}\, (n^{-1}\Sigma_{i=1}^n H(X_i)^2)^{1/2} / \bar{Y}_n|$$

which implies

$$E(\chi_n^{\ \rho}\delta_{2,n}^{\ \rho} \mid P_n) \leq E(\, |1 - \bar{Y}_n|^{\rho}\, (n^{-1}\Sigma_{i=1}^n Y_{i,n}^2)^{\rho/2}\, \bar{Y}_n^{\ -\rho}\,)$$

$$(n^{-1}\Sigma_{i=1}^n H(X_i)^2)^{1/2} \ .$$

Then, Hölder's inequality yields

$$E(\chi_n^{\ \rho}\delta_{2,n}^{\ \rho} \mid P_n) \leq (E|1 - \bar{Y}_n|^{3\rho})^{1/3}\, (E(n^{-1}\Sigma_{i=1}^n Y_{i,n}^2)^{3\rho/2})^{1/3}$$

$$(E|1/\bar{Y}_n|^{3\rho})^{1/3}\ (n^{-1}\Sigma_{i=1}^n H(X_i)^2)^{1/2}$$

which is less than

$$n^{-\rho/2}\ MZ((Y-1),3\rho)^{1/3}\ MZ(Y_{i,n}^2,3\rho/2)^{1/3}\ (E(n/S_n)^{3\rho})^{1/3} \tag{1.14}$$
$$(n^{-1}\Sigma_{i=1}^n H(X_i)^2)^{1/2}\ .$$

Then, (1.13) and (1.14) and (1.1) give Theorem II.2.1. □

Proof of (II.2.5) and (II.2.6) : Since we have $E\ H(X_i)^2 < \infty$ we take c large enough in Theorem II.2.1 so that the only terms which asymptotically remain are those with a sum of the form $n^{-1}\Sigma_{i=1}^n H(X_i)^2$ and $n^{-1}\Sigma_{i=1}^n H(X_i)^{2\rho/(\rho-1)}$. Then Marcinkiewicz-Zygmund's theorem gives the result. □

2) Proof of Proposition II.2.1

Let $\delta > 0$, and define

$$\delta_n := \delta\ n^{-1/2}(\log n)^{1/2}\ 128^{1/2}\ .$$

By Markov's inequality, we have

$$P\{\ |P_n h - Ph| \geq \delta_n/4\ \} \leq 16\ \delta_n^{-2}(\log n)^{-1} PH^2/128$$

Then, for n sufficiently large, the above upper bound is less than 1/2 and the symetrization steps in Pollard (1984) lead to

$$P\{\ d_{\mathcal{H}}(P_n,P) \geq \delta_n\ \} \leq 4\ P\{\ \sup_{h\in\mathcal{H}}|P_n^\circ h| \geq \delta_n/4\ \}\ ,$$

with $P_n^\circ := n^{-1}\Sigma_{i=1}^n\ \varepsilon_i\ \delta_{X_i}$ where the ε_i are define in section 1. But we have

$$P\{\ \sup_{h\in\mathcal{H}}|P_n^\circ h| \geq \delta_n/4\ |\ P_n\ \}$$

$$\le 2\ N(\delta_n/8, P_n, \mathcal{H})\ \max_j\ P\{\ |P_n^\circ h_j| \ge \delta_n/8 \mid P_n\ \}$$

$$\le 2\ N(\delta_n/8, P_n, \mathcal{H})\ \exp(-2^{-1} n\delta_n/(64\ P_n H^2)) \wedge 1$$

by using Hoeffding's inequality. Thus, using our polynomial assumption on the covering number $N(.,.,\mathcal{H})$, we get

$$P\{\ \sup_{h\in\mathcal{H}} |P_n h| \ge \delta_n/4\ \}$$

$$\le E_P((2\ P_n H/\delta_n)^D \exp(-2^{-1} n\delta_n/(64\ P_n H^2)) \wedge 1) \wedge 1$$

$$\le 2A\ (n/\log n)^{D/2}\ E\Big((P_n H)^D \exp(-\varepsilon^2(\log n)/P_n H^2) \wedge 1\Big)$$

$$\le 2A\ (n/\log n)^{D/2}\ E\Big((P_n H)^D \exp(-\varepsilon^2(\log n)/P_n H^2) \wedge 1\Big)\mathbb{I}\{\ P_n H^2 < c\ \}$$

$$+\ 2A\ (n/\log n)^{D/2}\ P\{\ P_n H^2 \ge c\ \}\ .$$

$$\le 2A\ (n/\log n)^{D/2}\ E(P_n H)^D \exp(-\varepsilon^2(\log n)/c) \tag{2.1}$$

$$+\ 2A\ (n/\log n)^{D/2}\ E|P_n H^2 - PH^2|^{D+2+\eta}\ (c-PH^2)^{-(D+2+\eta)}$$

for any $c > PH^2$ and $\eta > 0$. Let $\varepsilon^2 > (D/2) + 1 + c^2$. Then (2.1) is the generic term of a convergent series, and Borel-Cantelli's lemma ensures that

$$d_{\mathcal{H}}(P_n, P) = O(n^{-1/2}(\log n)^{1/2}) \quad \text{a.s.} \qquad \square$$

3) Proof of Proposition II.4.1

Recall the notation

$$P_{Y,n} := n^{-1}\ \Sigma_{i=1}^n\ Y_{i,n}\ \delta_{X_i}\ ,$$

so that, using Taylor's formula,

$$P_{W,n} - P_n = P_{Y,n} - P_n + (1-\overline{Y}_n)P_{Y,n} + (1-\overline{Y}_n)^2 P_{Y,n} - (1-\overline{Y}_n)^3 P_{Y,n} \tag{3.1}$$

$$+\ (1-\overline{Y}_n)^4 (1+\theta_n(1-\overline{Y}_n))^5\ P_{Y,n}$$

where $0 \le \theta_n \le 1$.

The proof consists in expanding $T(P_{W,n}) - T(P_n)$ as U-statistics of $Y_1, \ldots, Y_n$ plus a negligible term. Then, calling $\psi_{Y,n}(t)$ the characteristic function of this U-statistics, we use the standard technique. Namely, we expand $\psi_{Y,n}(t)$ and integrate it, to obtain the Edgeworth expansion.

First, to expand $T(P_{W,n}) - T(P_n)$ in using the differentiability of T at P , we rewrite it as

$$
\begin{aligned}
(3.2) \qquad T(P_{W,n}) - T(P_n) &= T(P_{W,n}) - T(P) + T(P) - T(P_n) \\
&= DT_P^1(P_{W,n}-P) + (1/2)\, D^2T_P(P_{W,n}-P)^2 + (1/6)\, D^3T_P(P_{W,n}-P)^3 \\
&\quad - DT_P^1(P_n-P) - (1/2)\, D^2T_P(P_n-P)^2 - (1/6)\, D^3T_P(P_n-P)^3 + R_{W,n}^{(3)} .
\end{aligned}
$$

From (3.1) we deduce

$$
\begin{aligned}
&DT_P^1(P_{W,n}-P) - DT_P^1(P_n-P) = DT_P^1(P_{W,n}-P_n) \\
&= DT_P^1(P_{Y,n}-P_n) - (\overline{Y}_n-1)\, DT_P^1\, P_n - (\overline{Y}_n-1)\, DT_P^1(P_{Y,n}-P_n) \\
&+ (\overline{Y}_n-1)^2 DT_P^1\, P_n + (\overline{Y}_n-1)^2 DT_P^1(P_{Y,n}-P_n) + O_P(n^{-2}) .
\end{aligned}
$$

Then, for the quadratic term,

$$
\begin{aligned}
&D^2T_P(P_{W,n}-P)^2 - D^2T_P(P_n-P)^2 \\
&= D^2T_P(P_{W,n}-P_n)^2 + 2\, D^2T_P(P_{W,n}-P_n)(P_n-P)
\end{aligned}
$$

and using (3.1), we have

$$
\begin{aligned}
&D^2T_P(P_{W,n}-P_n)^2 \\
&= D^2T_P(P_{Y,n}-P_n)^2 - 2(\bar{Y}_n-1)\, D^2T_P(P_{Y,n}-P_n)(P_n-P) \\
&- 2(\bar{Y}_n-1)\, D^2T_P(P_{Y,n}-P_n)^2 + O_P(n^{-2})
\end{aligned}
$$

and

$$D^2T_P(P_{W,n}-P)(P_n-P)$$

$$= D^2T_P(P_{Y,n}-P_n)(P_n-P) - (\bar{Y}_n-1)\, D^2T_P(P_n-P)^2$$

$$- (\bar{Y}_n-1)\, D^2T(P_{Y,n}-P_n)(P_n-P) + o_P(n^{-2}) .$$

Next, for the cubic term,

$$D^3T_P(P_{W,n}-P)^3 - D^3T_P(P_n-P)^3$$

$$= D^3T_P(P_{W,n}-P_n)^3 + 3\, D^3T_P(P_{W,n}-P_n)^2(P_n-P) + 3D^3T_P(P_{W,n}-P_n)(P_n-P)^2$$

$$= D^3T_P(P_{Y,n}-P_n)^3 + 3\, D^3T_P(P_{Y,n}-P_n)^2(P_n-P) + 3\, D^3T_P(P_{Y,n}-P_n)(P_n-P)^2$$

$$+ o_P(n^{-2}) \quad \text{as} \quad n\to\infty .$$

Putting all the pieces together, we get the expansion

$$T(P_{W,n}) - T(P_n) = n^{-1} \Sigma_{i=1}^n (Y_{i,n} - 1)\, \tau_i$$

$$+ 2^{-1} n^{-2} \Sigma_{i=1}^n \Sigma_{j=1}^n (Y_{i,n} - 1)(Y_{j,n} - 1)\, \tau_{i,j}$$

$$+ 6^{-1} n^{-3} \Sigma_{i=1}^n \Sigma_{j=1}^n \Sigma_{k=1}^n (Y_{i,n} - 1)(Y_{j,n} - 1)(Y_{k,n} - 1)\, \tau_{i,j,k}$$

$$+ \tilde{R}^{(3)}_{W,n} ,$$

where the τ_i , $\tau_{i,j}$ and $\tau_{i,j,k}$ are defined in (II.4.6) and $\tilde{R}^{(3)}_{W,n}$ is a remainder which is negligible in all the following calculations under suitable assumptions. Let $\psi_{Y,n}(t)$ be the characteristic function of $n^{1/2}(T(P_{W,n}) - T(P_n))$ conditioned on P_n :

$$\psi_{Y,n}(t) = E\Big(\exp\big(\underline{i}t\, n^{-1/2}\, \Sigma_{i=1}^n Y_n(i)\, \tau_i\big)$$

$$\exp\big(\underline{i}t\, 2^{-1} n^{-3/2}\, \Sigma_{i=1}^n \Sigma_{j=1}^n Y_n(i)\, Y_n(j)\, \tau_{i,j}\big)$$

$$\exp\big(\underline{i}t\, 6^{-1} n^{-5/2}\, \Sigma_{i=1}^n \Sigma_{j=1}^n \Sigma_{k=1}^n Y_n(i)\, Y_n(j)\, Y_n(k)\, \tau_{i,j,k}\big) \mid P_n \Big)$$

where $\underline{i}^2 = -1$ and $Y_n(i) := Y_{n,i} - 1$.

Using Taylor's formula,

$$\psi_{Y,n}(t) = E\Big(\Pi_n(t) \big(1 + \underline{i}t\, 2^{-1} n^{-3/2} \Sigma_{j=1}^n \Sigma_{k=1}^n\ Y_n(i)\, Y_n(j)\, \tau_{i,j}$$

$$- t^2\, 8^{-1} n^{-3} (\Sigma_{i=1}^n \Sigma_{j=1}^n Y_n(i)\, Y_n(j)\, \tau_{i,j})^2 + r_{1,n}(t)\big)$$

$$\big(1 + \underline{i}t\, 6^{-1} n^{-5/2} \Sigma_{i=1}^n \Sigma_{j=1}^n \Sigma_{k=1}^n Y_n(i)\, Y_n(j)\, Y_n(k)\, \tau_{i,j,k} +$$

$$r_{2,n}(t)\big) \mid P_n \Big)$$

where $r_{1,n}(t)$ and $r_{2,n}(t)$ are some remainders and

$$\Pi_n(t) := \exp\big(\underline{i}t\, n^{-1/2} \Sigma_{i=1}^n Y_n(i)\, \tau_i\big) .$$

We rewrite $\psi_{Y,n}(t)$ as

$$\psi_{Y,n}(t) = E_Y(\Pi_n(t)) + S_{1,n}(t) + S_{2,n}(t) + S_{3,n}(t) + r_{3,n}(t)$$

where E_Y denotes the expectation with respect to the $Y_{i,n}$'s conditionally on P_n with

$$S_{1,n}(t) := E_Y\Big(\Pi_n(t)\big(\underline{i}t\, 2^{-1} n^{-3/2} \Sigma_{i=1}^n \Sigma_{j=1}^n Y_n(i)\, Y_n(j)\, \tau_{i,j}\big)\Big)$$

$$S_{2,n}(t) := E_Y\Big(\Pi_n(t)\big(-\underline{i}t^2\, 8^{-1} n^{-3} \big(\Sigma_{i=1}^n \Sigma_{j=1}^n Y_n(i)\, Y_n(j)\, \tau_{i,j}\big)^2\big)\Big)$$

$$S_{3,n}(t) := E_Y\Big(\Pi_n(t)\big(\underline{i}t\, 6^{-1} n^{-5/2} \Sigma_{i=1}^n \Sigma_{j=1}^n \Sigma_{k=1}^n Y_n(i)\, Y_n(j)\, Y_n(k)$$

$$\tau_{i,j,k}\big)\Big)$$

and $r_{3,n}(t)$ is a remainder which depends on $r_{1,n}(t)$ and $r_{2,n}(t)$ such that, under suitable conditions,

$$(3.3) \qquad r_{3,n}(t) = o\big(t^{2+b} n^{-(1+b/2)} + t^{a+1} n^{-(1+a)} + t^2 n^{-3/2} + n^{-2} t^3\big) \quad \text{as } n \to \infty$$

for some $a,b \in [0,1]$ and uniformly in t (see Bickel, Götze and Van-Zwet (1986) for a justification, and Bertail (1992, Appendix 1 of Chapter 4, reproduced here in Appendix 8), for a complete proof).

From Zhidong and Lincheng's (1986) Theorem 2, we have uniformly in t,

(3.4) $$E_Y \Pi_n(t) = \exp(-t^2 2^{-1} \sigma_{Y,n}^2 \sigma_{\tau,n}^2) \Big(1 - \underline{i} t^3 6^{-1} n^{-1/2} \sigma_{Y,n}^3 \sigma_{\tau,n}^3 \overline{\tau_1^3} \beta_{Y,n}$$

$$+ t^4 24^{-1} n^{-1} (\sigma_{Y,n}^4 \sigma_{\tau,n}^4 \overline{\tau_1^4} K_{Y,n} - 3)$$

$$- t^6 72^{-1} n^{-1} \sigma_{Y,n}^6 \sigma_{\tau,n}^6 (\overline{\tau_1^3})^2 \beta_{Y,n}^2 \Big)$$

$$+ \exp(-t^2/2) p(t)\, O(n^{-3/2}) \quad \text{as} \quad n \to \infty$$

where $p(t)$ is a polynomial in t, whose coefficients depend on the moments of the weights.

Notation : For sake of simplicity, all the polynomials which can be neglected in the Edgeworth expansion up to $o(n^{-1})$ are denoted $p(t)$.

Consider now $S_{1,n}(t)$:

$$S_{1,n}(t) = \underline{i} t\, 2^{-1} n^{-3/2} \Sigma_{i=1}^n \Sigma_{\substack{j=1 \\ j \neq i}}^n E_Y\big(Y_n(i) \exp(\underline{i} t n^{-1/2} Y_n(i) \tau_i)\big)$$

$$E_Y\big(Y_n(j) \exp(\underline{i} t n^{-1/2} Y_n(j) \tau_j)\big) E_Y\big(\Pi_{n-2}^{i,j}(t)\big) \tau_{i,j}$$

$$+ \underline{i} t\, 2^{-1} n^{-3/2} \Sigma_{i=1}^n \tau_{i,i} E_Y\big(Y_n^2(i) \exp(\underline{i} t n^{-1/2} Y_n(i) \tau_i)\big) E_Y\big(\Pi_{n-1}^i(t)\big)$$

with

$$\Pi_{n-1}^i(t) := \exp\Big(\underline{i} t n^{-1/2} \Sigma_{\substack{k=1 \\ k \neq i}}^n Y_n(k) \tau_k\Big) \quad \text{and}$$

$$\Pi_{n-2}^{i,j}(t) := \exp\Big(\underline{i} t n^{-1/2} \Sigma_{\substack{k=1 \\ k \neq i,j}}^n Y_n(k) \tau_k\Big) .$$

We now approximate $\Pi_{n-1}^i(t)$ and $\Pi_{n-2}^{i,j}(t)$ by $\Pi_n(t)$ which of course does not depend on i and j . For this, expand $\Pi_{n-1}^i(t)$ and $\Pi_{n-2}^{i,j}(t)$ as in (3.4). Doing this, we have expansions for $E_Y\, \Pi_{n-1}^i(t)$ and $E_Y\, \Pi_{n-2}^{i,j}(t)$ similar to (3.4) but with $\sigma_{\tau,n}$ replaced respectively by $\sigma_{\tau,n}^i$ and $\sigma_{\tau,n}^{i,j}$ (with some obvious notations). One can evaluate the differences $\sigma_{\tau,n} - \sigma_{\tau,n}^i$ and $\sigma_{\tau,n} - \sigma_{\tau,n}^{i,j}$ and get

$$S_{1,n}(t) = E_Y\big(\Pi_n(t)\big)$$

$$\Big(\underline{i}t\ 2^{-1} n^{-3/2}\ \Sigma_{i=1}^n \Sigma_{\substack{j=1\\ j\neq i}}^n\ E_Y\big(Y_n(i)\ \exp(\underline{i}tn^{-1/2} Y_n(i)\ \tau_i)\big)$$

$$E_Y\big(Y_n(j)\ \exp(\underline{i}tn^{-1/2} Y_n(j)\ \tau_j)\big)\ \tau_{i,j}$$

$$+\ \underline{i}t\ 2^{-1} n^{-3/2}\ \Sigma_{i=1}^n \tau_{i,i}\ E_Y\big(Y_n^2(i)\ \exp(\underline{i}tn^{-1/2} Y_n(i)\ \tau_i)\big)\ \Big)$$

$$+\ t^3\ n^{-3/2}\ \exp(-t^2/8)\ O(1)$$

where the O(1)-term is uniform in the range $|t| = O(n^{1/6})$. Now, Taylor's formula gives

$$E_Y\big(Y_n(i)\ \exp(\underline{i}tn^{-1/2} Y_n(i)\ \tau_i)\big)$$

$$= E_Y\Big(Y_n(i)\big(1 + \underline{i}tn^{-1/2} Y_n(i)\ \tau_i\ -\ t^2\ 2^{-1} n^{-1} Y_n^2(i)\ \tau_i^2\ \big)\Big)\ +\ t^3 O(n^{-3/2})$$

$$= \underline{i}t\ n^{-1/2}\ \sigma_{Y,n}^2 \tau_i\ -\ t^2\ 2^{-1} n^{-1}\ \beta_{Y,n} \sigma_{Y,n}^3 \tau_i^2\ +\ t^3 O(n^{-3/2})$$

and

$$E_Y\big(Y_n(i)^2\ \exp(\underline{i}tn^{-1/2} Y_n(i)\ \tau_i)\big)$$

$$= E_Y\Big(Y_n(i)^2\ \big(1 + \underline{i}tn^{-1/2} Y_n(i)\ \tau_i\ -\ t^2\ 2^{-1} n^{-1} Y_n^2(i)\ \tau_i^2\ \big)\Big)$$

$$+\ t^3\ O(n^{-3/2})$$

$$= \sigma_{Y,n}^2\ +\ \underline{i}t\ n^{-1/2}\ \beta_{Y,n} \sigma_{Y,n}^3 \tau_i\ -\ t^2\ 2^{-1} n^{-1}\ \sigma_{Y,n}^2\ K_{Y,n}\ \tau_i^2$$

$$+\ t^3\ O(n^{-3/2})\ .$$

Combining these results, we obtain

$$S_{1,n}(t) = E_Y(\Pi_n(t))\ \Big(\ \underline{i}t\ 2^{-1} n^{-3/2}\ \Sigma_{i=1}^n \Sigma_{\substack{j=1\\ j\neq i}}^n\ \tau_{i,j}$$

$$\big(\underline{i}t\ n^{-1/2}\ \sigma_{Y,n}^2 \tau_i\ -\ t^2\ 2^{-1} n^{-1}\ \beta_{Y,n} \sigma_{Y,n}^3 \tau_i^2\ \big)$$

$$\Big(\underline{i}t\, n^{-1/2}\, \sigma_{Y,n}^{2} \tau_j - t^2\, 2^{-1} n^{-1}\, \beta_{Y,n} \sigma_{Y,n}^{3}\, \tau_j^2 \Big)$$

$$+ \underline{i}t\, 2^{-1} n^{-3/2}\, \Sigma_{i=1}^{n}\, \tau_{i,i} \Big(\sigma_{Y,n}^{2} + \underline{i}t\, n^{-1/2}\, \beta_{Y,n} \sigma_{Y,n}^{3} \tau_i$$

$$- t^2\, 2^{-1} n^{-1}\, \sigma_{Y,n}^{4}\, K_{Y,n}\, \tau_i^2 \Big)\Big]$$

$$+ O(n^{-3/2}\, t^3) + p(t) \exp(-t^2/8)\, o(n^{-1})$$

$$= E_Y(\Pi_n(t)) \Big(-\underline{i}t^3\, 2^{-1} n^{-5/2}\, \sigma_{Y,n}^{4}\, \Sigma_{i=1}^{n} \Sigma_{\substack{j=1\\ j\neq i}}^{n}\, \tau_i \tau_j \tau_{i,j}$$

$$- \underline{i}t^4\, 2^{-1} n^{-3}\, \sigma_{Y,n}^{5}\, \beta_{Y,n}\, \Sigma_{i=1}^{n} \Sigma_{\substack{j=1\\ j\neq i}}^{n}\, \tau_i^2 \tau_j \tau_{i,j}$$

$$+ \underline{i}t^5\, 2^{-1} n^{-7/2}\, \sigma_{Y,n}^{6}\, \beta_{Y,n}^{2}\, \Sigma_{i=1}^{n} \Sigma_{\substack{j=1\\ j\neq i}}^{n}\, \tau_i^2 \tau_j^2 \tau_{i,j}$$

$$+ \underline{i}t\, 2^{-1} n^{-3/2}\, \sigma_{Y,n}^{2}\, \Sigma_{i=1}^{n} \tau_{i,i} - t^2\, 2^{-1} n^{-2}\, \beta_{Y,n} \sigma_{Y,n}^{3}\, \Sigma_{i=1}^{n}\, \tau_i \tau_{i,i}$$

$$- t^4\, 4^{-1} n^{-5/2}\, K_{Y,n} \sigma_{Y,n}^{4}\, \Sigma_{i=1}^{n} \tau_i^2 \tau_{i,i} + O(n^{-3/2}\, t^3) \Big)$$

$$+ p(t) \exp(-t^2/8)\, o(n^{-1})\ .$$

$S_{2,n}(t)$ and $S_{3,n}(t)$ may be handled similarly, leading to (see Appendix 8 for analogue calculations)

$$\psi_{Y,n}(t) = E_Y\big(\Pi_n(t)\big) \Big(1 - \underline{i}t^3\, 2^{-1} n^{-5/2}\, \sigma_{Y,n}^{4}\, \Sigma_{i=1}^{n} \Sigma_{\substack{j=1\\ j\neq i}}^{n}\, \tau_i \tau_j \tau_{i,j}$$

$$- t^4\, 2^{-1} n^{-3}\, \sigma_{Y,n}^{5}\, \beta_{Y,n}\, \Sigma_{i=1}^{n}\, \Sigma_{\substack{j=1\\ j\neq i}}^{n}\, \tau_i^2 \tau_j \tau_{i,j}$$

$$+ \underline{i}t\, 2^{-1} n^{-3/2}\, \sigma_{Y,n}^{2}\, \Sigma_{i=1}^{n} \tau_{i,i} - t^2\, 2^{-1} n^{-2}\, \beta_{Y,n} \sigma_{Y,n}^{3}\, \Sigma_{i=1}^{n} \tau_i \tau_{i,i}$$

$$- t^2\, 8^{-1} n^{-3} \Big(t^2 n^{-1}\, \sigma_{Y,n}^{4}\, \Sigma_{i=1}^{n}\, \Sigma_{\substack{j=1\\ j\neq i}}^{n}\, \tau_i \tau_j \tau_{i,j} \Big)^2$$

$$- t^2\, 4^{-1} n^{-3} \Big(-t^2\, n^{-1}\, \sigma_{Y,n}^{6}\, \Sigma_{i=1}^{n}\, \Sigma_{\substack{j=1\\ j\neq i}}^{n}\, \tau_i \tau_j \tau_{i,j} \Sigma_{i=1}^{n} \tau_i \Big)$$

$$- t^2 \, 8^{-1} n^{-3} \left(\sigma_{Y,n}^2 \, \Sigma_{i=1}^n \tau_{i,i}\right)^2 \; - t^2 \, 4^{-1} n^{-3} \sigma_{Y,n}^4 \, \Sigma_{i=1}^n \, \Sigma_{\substack{j=1 \\ j\neq i}}^n \tau_{i,j}^2$$

$$+ \underline{i}t \, 6^{-1} n^{-5/2} \Big(-\underline{i}t^3 \, n^{-3/2} \, \sigma_{Y,n}^6 \, \Sigma_{i=1}^n \, \Sigma_{j=1}^n \, \Sigma_{k=1}^n \, \tau_{i,j,k} \tau_i \tau_j \tau_k \quad \text{i, j, k distinct}$$

$$+ \underline{i}t \, n^{-1/2} \, \sigma_{Y,n}^4 \, \Sigma_{i=1}^n \, \Sigma_{\substack{j=1 \\ j\neq i}}^n \tau_{i,j,j} \, \tau_i \, \Big) \Big)$$

$$+ O(t^2 \, n^{-3/2}) + p(t) \exp(-t^2/8) \, o(n^{-1}) \, .$$

Then, using approximation (3.4) we get

$$\psi_{Y,n}(t) = \exp\left(-t^2 \, 2^{-1} \sigma_{Y,n}^2 \, \sigma_{\tau,n}^2\right)$$

$$\Big(1 - \underline{i}t^3 \, 6^{-1} n^{-1/2} \, \beta_{Y,n} \sigma_{Y,n}^3 \, \sigma_{\tau,n}^3 \, \overline{\tau_1^3}$$

$$+ t^4 \, 24^{-1} n^{-1} \left(K_{Y,n} \, \sigma_{Y,n}^4 \, \sigma_{\tau,n}^4 \, \overline{\tau_1^4} - 3\right)$$

$$- t^6 \, 72^{-1} n^{-1} \, \beta_{Y,n}^2 \, \sigma_{Y,n}^6 \, \sigma_{\tau,n}^6 \, (\overline{\tau_1^3})^2$$

$$- \underline{i}t^3 \, 2^{-1} n^{-1/2} \sigma_{Y,n}^2 \, \sigma_{\tau,n}^3 \, \overline{\tau_1 \tau_2 \tau_{1,2}} \, \left(1 - it^3 \, 6^{-1} n^{-1/2} \beta_{Y,n} \sigma_{Y,n}^3 \sigma_{\tau,n}^3 \overline{\tau_1^3} \, \right)$$

$$+ t^4 \, 2^{-1} n^{-1} \, \beta_{Y,n} \, \sigma_{Y,n}^5 \, \sigma_{\tau,n}^4 \, \overline{\tau_1 \tau_2^2 \tau_{1,2}}$$

$$+ \underline{i}t \, 2^{-1} n^{-1/2} \, \sigma_{Y,n}^2 \, \sigma_{\tau,n} \, \overline{\tau_{1,1}} \, \left(1 - it^3 \, 6^{-1} n^{-1/2} \beta_{Y,n} \sigma_{Y,n}^3 \sigma_{\tau}^3 \, \overline{\tau_1^3}\right)$$

$$- t^2 \, 2^{-1} n^{-1} \, \beta_{Y,n} \sigma_{Y,n}^3 \sigma_{\tau,n}^2 \, \overline{\tau_1 \tau_{1,1}} \, - t^6 \, 8^{-1} n^{-1} \left(\sigma_{Y,n}^4 \sigma_{\tau,n}^3 \, \overline{\tau_1 \tau_2 \tau_{1,2}}\right)^2$$

$$+ t^4 \, 4^{-1} n^{-1} \, \sigma_{Y,n}^6 \sigma_{\tau,n}^4 \, \overline{\tau_{1,1} \tau_1 \tau_2 \tau_{1,2}} + t^2 \, n^{-1} \sigma_{Y,n}^6 \, \sigma_{\tau,n}^4 \, \overline{\tau_1 \tau_2 \tau_{1,3} \tau_{2,3}}$$

$$- t^2 \, 8^{-1} n^{-1} \left(\sigma_{Y,n}^2 \sigma_{\tau,n} \, \overline{\tau_{1,1}} \, \right)^2 - t^2 \, 4^{-1} n^{-1} \, \sigma_{Y,n}^4 \, \sigma_{\tau,n}^2 \, \overline{\tau_{1,2}^2}$$

$$+ t^4 \, 6^{-1} n^{-1} \sigma_{Y,n}^6 \, \sigma_{\tau,n}^4 \, \overline{\tau_1 \tau_2 \tau_3 \tau_{1,2,3}} \, - t^2 \, 2^{-1} n^{-1} \sigma_{Y,n}^4 \sigma_{\tau,n}^2 \, \overline{\tau_{1,1,2} \tau_2} \, \Big)$$

$$+ r_{Y,n}(t)$$

where $r_{Y,n}(t)$ is a remainder of order

$$r_{Y,n}(t) = O(t^{2+b}\, n^{-(1+b/2)} + t^{a+1} n^{-(1+a)} + t^3 n^{-2} + t^2\, n^{-3/2})$$

$$+ \; p(t)\, \exp(-t^2/8)\; o(n^{-1})$$

uniformly over the interval $t \in [0,n^{1/9}]$.

This yields exactly

$$\psi_{Y,n}(t) = \exp\left(-t^2\, 2^{-1} \sigma_{Y,n}^{2}\, \sigma_{\tau,n}^{2}\right)$$

$$\Big(1 - n^{-1/2}\big(\underline{i}t\; 2^{-1}\, K^{Y}_{1,1,n}(P) + \underline{i}t^3\; 6^{-1}\, K^{Y}_{1,3,n}(P)\big)$$

$$- \; n^{-1}\big(\, t^2\; 2^{-1}\, K^{Y}_{2,2,n}(P) - t^4\; 24^{-1}\, K^{Y}_{2,4,n}(P) + t^6\; 72^{-1} K^{Y}_{2,6,n}(P)\big)\Big)$$

$$+ \; r_{Y,n}(t)$$

where $K^{Y}_{1,i,n}(P)$, $i = 1$, 3 and $K^{Y}_{2,j,n}(P)$, $j = 2$, 4 , 6 are given by (II.4.9) .

This gives the behaviour of $\psi_{Y,n}(t)$ for central values of t (that is for $t \in [0,n^{\lambda}]$, for some $\lambda \le 1/9$). The next step should be the application of Esseen's smoothing lemma. However this is only possible if we check that

$$\text{(3.5)} \qquad \limsup_{n\to\infty} \int_{n^{\lambda}}^{n\log n} |\psi_{Y,n}(t)|/t \; dt = 0.$$

In the non-bootstrap case, for U or V-statistics, this is achieved by a conditioning argument and some continuous property of the r.v.'s (see for instance Callaert, Janssen and Veraverbeke (1980)). In our case the calculations would be extremely tedious, but would follow their proof. Since our aim is essentially to give the form of the Edgeworth expansion without controlling all the remainders, we assume that (3.5) is satisfied. Under such assumption we may derive the Edgeworth expansion given in Proposition II.4.1 : invert $\psi_{Y,n}(t)$ to get $\mathcal{E}^{W}_{2,H}$ as defined in Proposition II.4.1. □

4) Proof of (II.5.6)

The main tool is the Edgeworth expansion given in Proposition II.4.1. We have

$$P\{\, n^{1/2}(T(P_{W,n})-T(P_n))/(\sigma_{W,n}S_n) \le x \mid P_n \,\}$$

(4.1) $$= P\{\, n^{1/2}(T(P_{W,n})-T(P_n))/(\sigma_{Y,n}\sigma_{\tau,n}) \le x(S_n/\sigma_{\tau,n})(\sigma_{W,n}/\sigma_{Y,n}) \mid P_n \,\}$$

Proposition II.4.1 provides an Edgeworth expansion for $n^{1/2}(T(P_{W,n}) - T(P_n))/(\sigma_{Y,n}\sigma_{\tau,n})$. We show in the following that $(S_n/\sigma_{\tau,n})(\sigma_{W,n}/\sigma_{Y,n})$ may be approximated by a deterministic quantity.

We first expand S_n^2 using the fact that (see (IV.4.4))

$$S_n^2 = E_W(\, n^{1/2}(T(P_{W,n})-T(P_n))/\sigma_{W,n}\,)^2$$

$$= E_W\Big(Q_{W,n} + 6^{-1}n^{-5/2}\,\Sigma_{i=1}^n \Sigma_{j=1}^n \Sigma_{k=1}^n (W_i^{(n)}W_j^{(n)}W_k^{(n)}-1)T_3(i,j,k)/\sigma_{W,n}$$

$$+ R^{(3)}(P_n,P) + R^{(3)}(P_{W,n},P_n) \Big)^2 .$$

We easily deduce from Appendix 3, and using that $E_W W_n(1)W_n(2) = -\sigma_{W,n}^2/(n-1)$ (see Appendix 1) :

(4.2) $$S_n^2 = n^{-1}\,\Sigma_{i=1}^n T_1(i)^2 + 2n^{-2}\,\Sigma_{i=1}^n \Sigma_{j=1}^n T_1(i)\,T_2(i,j)$$

$$+ \sigma_{W,n}^2(1/4n)\,(E_PT_2(X,X))^2 + \sigma_{W,n}^2\,(1/2n)\,E_PT_2(X,Y)^2$$

$$+ \sigma_{W,n}\beta_{W,n}\,n^{-1}\,E_PT_1(X)T_2(X,X) - n^{-2}\,\Sigma_{i=1}^n \Sigma_{\substack{j=1\\ j\ne i}}^n T_1(i)\,T_1(j)$$

$$+ \sigma_{W,n}^2\,n^{-1}\,E_PT_1(X)T_3(X,Y,Y) + n^{-2}\,\Sigma_{i=1}^n \Sigma_{j=1}^n E_PT_1(X)\,T_3(X,i,j)$$

$$+ n^{-2}\,\Sigma_{i=1}^n \Sigma_{j=1}^n E_PT_2(X,i)\,T_2(X,j) + o_p(n^{-1}) .$$

Consider now $\sigma_{\tau,n}^2$. From (II.4.6) we get, after a quick calculation

$$\sigma_{\tau,n}^2 = n^{-1}\,\Sigma_{i=1}^n T_1(i)^2 + n^{-2}\,\Sigma_{i=1}^n \Sigma_{j=1}^n E_PT_2(X,i)\,T_2(X,j)$$

$$+ 2\,n^{-2}\,\Sigma_{i=1}^n \Sigma_{j=1}^n T_1(i)\,T_2(i,j)$$

$$+ n^{-2}\,\Sigma_{i=1}^n \Sigma_{j=1}^n E_PT_1(X)\,T_3(X,i,j)$$

$$- n^{-2} \Sigma_{i=1}^{n} \Sigma_{j=1}^{n} T_1(i)\, T_1(j) + o(n^{-1}) .$$

We see that

$$S_n^2 - \sigma_{\tau,n}^2 = \sigma_{W,n}^2 (1/4n)\, (E_P T_2(X,X))^2 + \sigma_{W,n}^2 (1/2n)\, E_P T_2(X,Y)^2$$

$$+ \sigma_{W,n} \beta_{W,n}\, n^{-1}\, E_P T_1(X) T_2(X,X) + \sigma_{W,n}^2\, n^{-1}\, E_P T_1(X) T_3(X,Y,Y)$$

$$+ n^{-1} S^2 + o_P(n^{-1}) \quad \text{as} \quad n \to \infty$$

(compare with the expression of $K_{2,2,n}^{Y}(P)$ in (II.4.9)).

Next,

$$S_n/\sigma_{\tau,n} = (1 + S^{-2}(S_n^2 - S^2))^{1/2}\, (1 + S^{-2}(\sigma_{\tau,n}^2 - S^2))^{-1/2}$$

$$= 1 + (1/2)\, S^{-2}\, (S_n^2 - \sigma_{\tau,n}^2) + (\sigma_{\tau,n}^2 - S_n^2) O(n^{-1/2})$$

$$= 1 + (1/2)\, S^{-2}\, (S_n^2 - \sigma_{\tau,n}^2) + o(n^{-1})$$

$$= 1 + (1/2n)\, K_{2,2,n}^{Y}(P) - (1/2n)\Delta_{Y,n} + o(n^{-1}) . \tag{4.3}$$

where $\Delta_{Y,n}$ is defined in (II.5.9′). It follows from (4.1) and (4.3) and Proposition II.4.1 that

$$P\{\, n^{1/2}(T(P_{W,n}) - T(P_n))/(S_n \sigma_{W,n}) \le x \mid P_n \,\}$$

$$= \mathcal{E}_{2,H}^{W}(x) + (1/2n)\, K_{2,2,n}^{Y}(P)\, x\, \phi(x) - n^{-1}\Delta_{Y,n}/2\; x\, \phi(x) \tag{4.4}$$

$$+ (\sigma_{W,n}/\sigma_{Y,n} - 1)\, x\, \phi(x) + o_P(n^{-1}) .$$

Finally, (4.4) proves (II.5.6) since we have

$$\sigma_{W,n}^2 = \sigma_{Y,n}^2 + n^{-1}(3\sigma_{Y,n}^4 - \sigma_{Y,n}^2 - 2\sigma_{Y,n}^3\, \beta_{Y,n})(1+o(1)) \quad \text{and}$$

$$\beta_{W,n} = \beta_{Y,n} + O(n^{-1}) \quad \text{as} \quad n \to \infty .$$

(see Appendix I for these last relations. □

5) Proof of (II.5.9)

The first step is to obtain a stochastic expansion for $S_n^{\,2}$. The result will follow from the Edgeworth expansion of $H_n(x,P)$ (see II.4.1) Equation (4.2) leads to

$$S_n^{\,2} = \tilde{S}^2(P_n) + n^{-1}\kappa_{Y,n} + o_P(n^{-1}) \quad \text{with}$$

$$\tilde{S}^2(Q) := S^2 + \int S^{(1)}(x,P)d(Q-P)(x)$$

$$+ 2^{-1} \iint S^{(2)}(x,y,P)\ d(Q-P)(x)\ d(Q-P)(y) \quad \text{and}$$

$$\kappa_{Y,n} := (\sigma_{W,n}^{\;4}/4)\ (E\ T_2(X,X))^2 + \sigma_{W,n}\beta_{W,n}\ E_PT_1(X)T_2(X,X)$$

$$+ \sigma_{W,n}^{\;2}\ (1/2)E_PT_2(X,Y)^2 + \sigma_{W,n}^{\;2}\ E_PT_1(X)T_3(X,Y,Y) + S^2 .$$

Notice that

$$\kappa_{Y,n} = S^2\ K^Y_{2,2,n}(P) - \Delta_{Y,n}\ S^2 + o_p(1) \quad \text{as} \quad n\to\infty .$$

We rewrite $K_n(x,P)$ as

$$K_n(x,P) = P\{n^{1/2}(T(P_n)-T(P))/(\tilde{S}^2(P_n)+n^{-1}\kappa_{Y,n}+o_p(n^{-1}))^{1/2} \le x \mid P_n\}$$

$$= P\{\ n^{1/2}(T(P_n)-T(P))/\tilde{S}(P_n) \le x(1 + \kappa_{Y,n}/(2nS^2) + o_P(n^{-1}))\ \mid P_n\ \}$$

$$= P\{\ n^{1/2}(T(P_n)-T(P))/\tilde{S}(P_n) \le x\ \} + x\phi(x)\ \kappa_{Y,n}/(2nS^2) + o(n^{-1}) ,$$

under suitable regularity conditions to control all the remainder terms. Therefore, it remains for us to obtain an Edgeworth expansion for

$$\tilde{K}_n(x,P) := P\{\ n^{1/2}(T(P_n)-T(P))/\tilde{S}(P_n) \le x\ \} .$$

Consider the functional

$$f(Q) = (T(Q) - T(P))/\tilde{S}(Q) .$$

Clearly, $f(P) = 0$ and

$$f^{(1)}(x,P) = T^{(1)}(x,P)/S(P) .$$

Therefore,

$$\tilde{K}_n(x,P) := P\{ n^{1/2}(f(P_n)-f(P))/(E\, f^{(1)}(X,P)^2)^{1/2} \le x \} .$$

Now, we see that if we replace the functional $f(.)$ by $T(.)$, $\tilde{K}_n(x,P)$ is exactly $H_n(x,P)$ defined in (II.4.1). Hence, the Edgeworth expansion for $\tilde{K}_n(x,P)$ is that of $H_n(x,P)$ when we replace $T(.)$ by $f(.)$. We replace the influence function $T^{(1)}$ and $T^{(2)}$, $T^{(3)}$ in the expressions of $K_{1,1}(P)$, $K_{1,3}(P)$, $K_{2,2}(P)$, $K_{2,4}(P)$ in (II.4.3) by the influence function $f^{(1)}$, $f^{(2)}$ and $f^{(3)}$ to get the Edgeworth expansion for $\tilde{K}_n$ and (II.5.9) (see Appendix 7 for the calculation of $f^{(1)}$, $f^{(2)}$ and $f^{(3)}$) □

6) <u>Proof of (II.5.10)</u>

We start with the following key lemma which shows how to modify an Edgeworth expansion when a statistic is slightly altered.

<u>Lemma 6.1</u> : **Let $\Pi_1(P)$ and $\Pi_2(P)$ be two functionals respectively third and first order differentiable, under suitable assumptions. Assume that $\Pi_1(P_n)$ admits an Edgeworth expansion up to $o(n^{-1})$, given by**

$$P\{ n^{1/2}(\Pi_1(P_n) - \Pi_1(P)) / E_P\, \Pi_1^{(1)}(X,P)^2 \le x \} = \mathcal{E}_{\Pi,2}(x) + o(n^{-1}).$$

Then, under the Cramér condition

$$\limsup_{t\to\infty} |E \exp(\underline{i}t(\Pi_1^{(1)}(X,P) + n^{-1}\Pi_1^{(2)}(X,P)))| < 1 ,$$

the statistic $\Pi_1(P_n) + n^{-1}\Pi_2(P_n)$ correctly standardized admits an Edgeworth expansion

$$P\{ n^{1/2} (\Pi_1(P_n) + n^{-1}\Pi_2(P_n) - \Pi_1(P) - n^{-1}\Pi_2(P)) / \Xi_n(P) \le x \}$$

$$= \mathcal{E}_{\Pi,2}(x) + o(n^{-1}) \quad \text{as } n\to\infty ,$$

where

$$\Xi_n^{\,2}(P) := E_P\Pi_1^{(1)}(X,P)^2 + 2n^{-1}\, E_P\Pi_1^{(1)}(X,P)\ \Pi_2^{(1)}(X,P)\ .$$

Proof of Lemma 6.1 : Expanding $\Pi_1(P_n)$ up to its third order derivative and $\Pi_2(P_n)$ up to its first order derivative, we get

$$n^{1/2}\ (\Pi_1(P_n) + n^{-1}\Pi_2(P_n) - \Pi_1(P) - n^{-1}\Pi_2(P))$$

$$= n^{1/2}\left(n^{-1}\ \Sigma_{i=1}^{n}\big(\Pi_1^{(1)}(X_i,P) + n^{-1}\Pi_2^{(1)}(X_i,P)\ \big)\right.$$

$$+ (1/2n^2)\ \Sigma_{i=1}^{n}\ \Sigma_{j=1}^{n}\ \Pi_1^{(2)}(X_i,X_j,P)$$

$$\left. + (1/6n^3)\ \Sigma_{i=1}^{n}\ \Sigma_{j=1}^{n}\ \Sigma_{k=1}^{n}\Pi_1^{(3)}(X_i,X_j,X_k,P) + R_n \right)$$

where the remainder R_n may be controlled under suitable conditions (see chapter I for some exemples).

Thus, under the Cramér condition, the Edgeworth expansion of $\Pi_1(P_n)$ + $n^{-1}\Pi_2(P_n)$ is given by a general functional T with first gradient $T^{(1)}(x,P)$ replaced by $\Pi_1^{(1)}(x,P) + n^{-1}\Pi_2^{(1)}(x,P)$. Since T involved in (II.4.2) may depend on n (see the arguments of Bickel, Götze and Van Zwet (1986) and Bertail (1992)), it suffices to calculate the coefficients in the Edgeworth expansion.

First,

$$\Xi^2_{\ n}(P) = E_P(\Pi_1^{(1)}(X,P) + n^{-1}\Pi_2^{(1)}(X,P))^2 + o(n^{-1})$$

$$= E_P\Pi_1^{(1)}(X,P)^2 + 2n^{-1}\ E\ \Pi_1^{(1)}(X,P)\Pi_2^{(1)}(X,P) + .$$

Moreover, the terms given in (II.4.3) calculated with $\Pi_1^{(1)}(.,P)$ + $n^{-1}\Pi_2^{(1)}(.,P)$ are equal to those calculated with $\Pi_1^{(1)}(.,P)$ plus a term of order $o(n^{-1})$, leading to the result of the lemma. □

The general pattern of the proof of (II.5.10) is to introduce suitable functionals of the empirical p.m. P_n for which we can obtain an Edgeworth expansion. We now turn to the proof of (II.5.10). We first start in rewriting the approximation of $\gamma_{W,n}(\alpha)$ given in (II.5.8). For this, define the functionals

$$\tau^{(1)}(x,Q) := T^{(1)}(x,P) + E_Q T^{(2)}(x,Y,P) + (1/2)E_Q T^{(3)}(x,Y,Z,P)$$

$$- E_Q T^{(1)}(Y,P) - E_Q T^{(2)}(Y,Z,P) ,$$

$$\tau^{(2)}(x,y,Q) := T^{(2)}(x,y,P) + E_Q T^{(3)}(x,y,Z,P)$$

$$- 2 (E_Q T^{(2)}(x,Z,P) + E_Q T^{(2)}(y,Z,P)) - T^{(1)}(x,P) - T^{(1)}(y,P)$$

$$+ 2 E_Q T^{(1)}(Z,P) ,$$

$$\tau^{(3)}(x,y,z,Q) := T^{(3)}(x,y,z,P)$$

$$+ 2 (T^{(1)}(x,P) + T^{(1)}(y,P) + T^{(1)}(z,P))$$

$$- 2(T^{(2)}(x,y,P) + T^{(2)}(x,z,P) + T^{(2)}(y,z,P)) ,$$

$$\sigma_{\tau,n}^2(Q) := E_Q \tau^{(1)}(X,Q)^2 ,$$

$$K_{1,1}^{Y,n}(Q) := \sigma_{Y,n} E_Q \tau^{(2)}(X,X,Q)/\sigma_\tau(Q) ,$$

$$K_{1,3}^{Y,n}(Q) := \beta_{Y,n} E_Q \tau^{(1)}(X,Q)^3/\sigma_\tau(Q)^3$$

$$+ 3 \sigma_{Y,n} E_Q \tau^{(1)}(X,Q)\tau^{(1)}(Y,Q)\tau^{(2)}(X,Y,Q)/\sigma_\tau(Q)^3 ,$$

$$K_{2,2}^{Y,n}(Q) := 4^{-1} \sigma_{Y,n}^2 (E_Q \tau^{(2)}(X,X,Q)/\sigma_\tau(Q))^2$$

$$+ 2^{-1} \sigma_{Y,n}^2 E_Q \tau^{(2)}(X,Y,Q)^2/\sigma_\tau(Q)^2$$

$$+ \sigma_{Y,n}^2 E_Q \tau^{(1)}(X,Q)\tau^{(3)}(X,Y,Y,Q)/\sigma_\tau(Q)^2$$

$$+ \sigma_{Y,n}\beta_{Y,n} E_Q \tau^{(1)}(X,Q)\tau^{(2)}(X,X,Q)/\sigma_\tau(Q)^2 ,$$

$$K_{2,4}^{Y,n}(Q) := 2 K_{1,1}^{Y,n}(Q) K_{1,3}^{Y,n}(Q) + K_{Y,n} E_Q \tau^{(1)}(X,Q)^4/\sigma_\tau(Q)^4 - 3$$

$$+ 12 \beta_{Y,n} \sigma_{Y,n} E_Q \tau^{(1)}(X,Q)\tau^{(1)}(Y,Q)^2\tau^{(2)}(X,Y,Q)/\sigma_\tau(Q)^4$$

$$+ 12\ \sigma_{Y,n}^{2}\ E_Q\tau^{(1)}(X,Q)\tau^{(1)}(Y,Q)\tau^{(2)}(X,Z,Q)\tau^{(2)}(Y,Z,Q)/\sigma_\tau(Q)^4$$

$$+ 4\ \sigma_{Y,n}^{2}\ E_Q\tau^{(1)}(X,Q)\tau^{(1)}(Y,Q)\tau^{(1)}(Z,Q)\tau^{(3)}(X,Y,Z,Q)/\sigma_\tau(Q)^4 ,$$

$$K_{2,6}^{Y,n}(Q) := K_{1,3}^{Y,n}(Q)^2 ,$$

so that $K_{i,j,n}^{Y}(P) = K_{i,j}^{Y,n}(P_n)$ ($i = 1, 2$ and $j = 1, 2, 3, 4, 6$). These functionals are continuous at P and

$$(6.1) \qquad K_{i,j,n}^{Y}(P) = K_{i,j}^{Y,n}(P) + o_P(1) \quad \text{as} \quad n\to\infty .$$

From (6.1) and (II.5.7)-(II.5.8), we deduce that

$$\gamma_{W,n}(\alpha) = c'_{Y,n}(\alpha) + o_p(n^{-1}) \quad \text{with}$$

$$c'_{Y,n}(\alpha) := u_\alpha + n^{-1/2}\ (\ 2^{-1}K_{1,1}^{Y,n}(P_n) + 6^{-1}K_{1,3}^{Y,n}(P_n)(u_\alpha^{\ 2}-1)\ )$$

$$+ n^{-1}\ 24^{-1}K_{2,4}^{Y,n}(P)(u_\alpha^{\ 3}-3u_\alpha) + n^{-1}\ 72^{-1}K_{2,6}^{Y,n}(P)(u_\alpha^{\ 5}-10u_\alpha^{\ 3}+15u_\alpha)$$

$$+ n^{-1} \left(3^{-1}K_{1,3}^{Y,n}(P) - 4^{-1}K_{1,1}^{Y,n}(P) - 12^{-1}K_{1,3}^{Y,n}(P)(u_\alpha^{\ 2}-1) \right)$$

$$\left(2^{-1}K_{1,1}^{Y,n}(P) + 6^{-1}K_{1,3}^{Y,n}(P)(u_\alpha^{\ 2}-1) \right) u_\alpha$$

which is the desired approximation.

It is rather simple, thought it is not our purpose to find conditions such that

$$(6.2) \qquad P\{\ n^{1/2}(T(P_n)-T(P))/S \le \gamma_{n,W}(\alpha)\ \} = p_{c,n}^{H,\prime}(\alpha) + o(n^{-1})$$

where

$$p_{c,n}^{H,\prime}(\alpha) := P\{\ n^{1/2}(T(P_n)-T(P))/S \le c'_{Y,n}(\alpha)\ \}$$

$$(6.3) \qquad = P\{\ n^{1/2}(T(P_n)-T(P))/S - n^{-1/2}(\ (1/2)(K_{1,1}^{Y,n}(P_n) - K_{1,1}^{Y,n}(P))$$

$$+ 6^{-1}(K_{1,3}^{Y,n}(P_n) - K_{1,3}^{Y,n}(P))(u_\alpha^2-1)\ ) \le c''_{Y,n}(\alpha)\ \}$$

and

$$c''_{Y,n}(\alpha) := u_\alpha + n^{-1/2}\ (\ 2^{-1}K_{1,1}^{Y,n}(P) + 6^{-1}K_{1,3}^{Y,n}(P)(u_\alpha^{\ 2}-1)\ )$$

$$+ n^{-1}\, 24^{-1} K_{2,4}^{Y,n}(P)(u_\alpha^{\;3}-3u_\alpha) + n^{-1}\, 72^{-1} K_{2,6}^{Y,n}(P)(u_\alpha^{\;5}-10u_\alpha^{\;3}+15u_\alpha)$$

$$+ n^{-1}\Big(3^{-1} K_{1,3}^{Y,n}(P) - 4^{-1} K_{1,1}^{Y,n}(P) - 12^{-1} K_{1,3}^{Y,n}(P)(u_\alpha^{\;2}-1) \Big)$$

$$\Big(2^{-1} K_{1,1}^{Y,n}(P) + 6^{-1} K_{1,3}^{Y,n}(P)(u_\alpha^{\;2}-1) \Big)\, u_\alpha \,.$$

To calculate the Edgeworth expansion of (6.3), we can consider the functionals

$$\Pi_1(Q) := T(Q)/S(P) \quad \text{and}$$

$$\Pi_2(Q) := -\Pi_2^{Y,n}(Q) := -(1/2)K_{1,1}^{Y,n}(Q) - (1/6)K_{1,3}^{Y,n}(Q)(u_\alpha^{\;2}-1)$$

and apply Lemma 6.1. It is easy to calculate the influence function $\Pi_2^{Y,n(1)}(x,P)$ of $\Pi_2^{Y,n}$ at P which is given in section II.5.

Consequently, with notations of Lemma 6.1 ,

$$\Xi_n^{\;2}(P) = 1 - 2n^{-1}\, E\, T^{(1)}(X,P)\Pi_2^{Y,n(1)}(X,P)/S(P) \,.$$

Apply Lemma 6.1 to get

$$P\{\, n^{1/2}(T(P_n)-T(P))/S - n^{-1/2}(\, (1/2)(K_{1,1}^{Y,n}(P_n) - K_{1,1}^{Y,n}(P))$$

$$+ (1/6)(K_{1,3}^{Y,n}(P_n) - K_{1,3}^{Y,n}(P))(u_\alpha^2-1)\,) \le x \,\}$$

$$= P\{\, n^{1/2}\, (\Pi_1(P_n) + n^{-1}\Pi_2(P_n) - \Pi_1(P) - n^{-1}\Pi_2(P)) \,/\, \Xi_n(P)$$

$$\le x/\Xi_n(P) \,\}$$

$$= H_n(x/\Xi_n(P)) + o(n^{-1})$$

(6.4) $$= H_n(x) + (1/nS(P))\, E\, T^{(1)}(X,P)\Pi_2^{Y,n(1)}(X,P)\; x\phi(x) + o(n^{-1}) \,.$$

Finally, combine (6.2), (6.3) and (6.4) to obtain (II.5.10). □

7) Proof of (II.5.11)

The proof of (II.5.11) follows from that of (II.5.10). It suffices to replace the functional $\Pi_1(Q)$ by $f(Q)$ defined in the proof of (II.5.9). Then, in equation (6.4) of the proof of (II.5.9), we replace $H_n(x,P)$ by $K_n(x,P)$, and in the application of Lemma 6.1 in the proof of (II.5.10), we replace $\Pi_1^{(1)}(x,P)$ by $f^{(1)}(x,P)$ (but they have the common value $T^{(1)}(x,P)/S(P)$). □

8) Proof of Theorem II.6.2

Repeating the arguments leading to Corollary 6.2.3 of Dembo and Zeitouni (1993), we see that $P_{Y,n}$ obeys a large deviation principle with rate function the convex conjugate of

$$\Lambda(f) := \lim_{n\to\infty} n^{-1} \log E\left(\exp\left(\Sigma_{i=1}^n Y_i f(X_i) \right) \Big| X_1,\dots,X_n\right)$$

$$= \lim_{n\to\infty} P_n \log L\circ f$$

$$= P \log L\circ f \quad P\text{-a.s.}$$

where f varies in the set $\mathcal{C}_b(B)$ of all continuous and bounded functions on B (notice that the last equality above does not require $\log L\circ f$ to be P-integrable; assumption (II.6.4) ensures that $P \log L\circ f$ is finite or equal to $+\infty$. In this last case, the equality follows from a usual truncature argument : replace $\log L\circ f$ by $\min(\log L\circ f, M)$, prove the equality, and then let M tend to infinity).

Hence, the rate function for the large deviation principle is

$$\Lambda^*(Q) := \sup_{f\in\mathcal{C}_b(B)} Qf - P \log L\circ f .$$

We now identify $\Lambda^*(.)$.

Assume first that Q is not absolutely continuous w.r.t. P, and let A be a Borel set in B such that $Q(A) > 0 = P(A)$. Set $\tilde f_c = c\mathbb{1}_A$ for any $c > 0$. Approximating $\tilde f_c$ by continuous functions, we obtain that

$$\Lambda^*(Q) \ge Q\tilde{f}_c - P \log L\circ\tilde{f}_c = c\, Q(A)\ .$$

Let c tend to infinity to conclude that in this case $\Lambda^*(Q) = \infty$.

Assume from now on, that $Q \ll P$ and let $\phi = dQ/dP$. Then

$$\sup_{f\in\mathcal{C}_b(B)} Qf - P \log L\circ f \le P \sup_{f\in\mathcal{C}_b(B)} f\phi - \log L\circ f$$

$$= P\ \gamma\circ\phi = J(Q,P)\ .$$

In order to conclude that $J(Q,P) = \Lambda^*(Q)$, let $g = m^{\leftarrow}\circ\phi$. The function g is well defined thanks to assumption (II.6.5). Then

$$Qg - P \log L\circ g = P\,(\ \phi\, m^{\leftarrow}\circ\phi - \log L\circ m^{\leftarrow}\circ\phi\ ) = P\ \gamma\circ\phi\ .$$

Approximating g by continuous and bounded function, it follows that $\Lambda^*(Q) \ge J(Q,P)$, which completes the proof.

CHAPTER VI

PROOFS OF RESULTS OF CHAPTER III

1) Proof of Theorem III.2.1

The proof consists mainly in a reduction to usual uniform empirical process and an application of the weighted approximation of M. Csörgö, S. Csörgö, Horváth and Mason (1986) or Mason and Van Zwet (1987).

Let $\mathbb{U}_n$ be the empirical d.f. of U_1, ..., U_n defined by

$$\mathbb{U}_n(x) = n^{-1} \Sigma_{i=1}^n \mathbb{I}\{ U_i \le x \} , \quad 0 \le x \le 1 ,$$

and let

$$\alpha_n(x) := n^{1/2}(\mathbb{U}_n(x) - x) \quad , 0 \le x \le 1 .$$

We can assume that the U_i 's are defined on the probability space of Mason and Van Zwet (1987) so that there exists a sequence of Brownian bridges $(B_n)_{n\ge1}$ such that, for any $0 \le \nu \le 1/2$,

$$(1.1) \qquad n^{\nu}\left\| \frac{\alpha_n - B_n}{(I(1-I))^{1/2-\nu}} \right\|_n = O_P(1) , \quad \text{as} \quad n\to\infty .$$

Then we have $\mathbb{D}_n = \mathbb{U}_n \circ g_n$ and

$$(1.2) \qquad n^{\nu} \frac{n^{1/2}(\mathbb{D}_n - \tilde{g}_n)(x) - B_n(\tilde{g}_n(x)) - n^{1/2}(g_n(x)-\tilde{g}_n(x))}{(x(1-x))^{1/2-\nu}}$$

$$= n^{\nu} \frac{(\alpha_n(g_n(x))-B_n(g_n(x)))}{(g_n(x)(1-g_n(x)))^{1/2-\nu}} \frac{(g_n(x)(1-g_n(x)))^{1/2-\nu}}{(x(1-x))^{1/2-\nu}}$$

$$+ n^{\nu} \frac{B_n(\tilde{g}_n(x)) - B_n(g_n(x))}{(x(1-x))^{1/2-\nu}}$$

$$=: \delta_{1,n}(x) + \delta_{2,n}(x) .$$

Using (1.1) and (III.2.6), we obtain that

(1.3) $$\| \delta_{1,n}(x) \|_n = O_p(1) \ , \text{ as } n\to\infty \ .$$

Next, let Δ_n be the left hand side of (III.2.5), i.e.

$$\Delta_n := \sup_{t\in[1/n,1-1/n]} \sup_{|s|\le 1/(2n)} n^{1/2} \frac{|g_n(t+s)-\tilde{g}_n(t)|}{(t(1-t))^{1/2}} \ .$$

From (III.2.5) we deduce that for any $\varepsilon > 0$ and any λ large enough

$$P\{ \|\delta_{2,n}(x)\|_n > u \}$$

(1.4) $$\le P\{ \|\delta_{2,n}(x)\|_n > u \ , \ \Delta_n \le \lambda \} + \varepsilon \ .$$

We have

$$\|\delta_{2,n}(x)\|_n \le \max_{1\le i\le n} \sup_{|s-i/n|\le 1/2n} \frac{|B_n(g_n(s))-B_n(\tilde{g}_n(i/n))|}{(s(1-s))^{1/2-\nu}}$$

$$+ \max_{1\le i\le n} \sup_{|s-i/n|\le 1/2n} \frac{|B_n(\tilde{g}_n(s))-B_n(\tilde{g}_n(i/n))|}{(s(1-s))^{1/2-\nu}} \ .$$

Using the same technique as in Mason (1991), we obtain

$$P\left(\sup_{|x-i/n|\le 1/2n} n^{1/2} \frac{|B_n(g_n(x))-B_n(\tilde{g}_n(i/n))|}{i^{1/2-\nu}} \ge u \ , \ \Delta_n \le \lambda \right)$$

(1.5) $$\le P\left(\sup_{|t-\tilde{g}_n(i/n)|\le\mu_{i,n}} |B_n(\tilde{g}_n(i/n)+t)-B_n(\tilde{g}_n(i/n))| \ge n^{1/2} u \ i^{1/2-\nu} \right.$$

$$\left. \Delta_n \le \lambda \right)$$

for any $\mu_{i,n}$ such that

$$\mu_{i,n} \ge \sup_{|s-i/n|\le 1/2n} | g_n(s) - \tilde{g}_n(i/n) | \ .$$

Clearly, when $\Delta_n \le \lambda$, using (III.2.5) we can take

$$\mu_{i,n} = \lambda \ i^{1/2} / \ n \ .$$

Then we can upper-bound (1.5) by

$$(1.6) \qquad P\Big(\sup_{|t-\tilde{g}_n(i/n)| \le \mu_{i,n}} |B_n(\tilde{g}_n(i/n)+t) - B_n(\tilde{g}_n(i/n))| \ge u\, \mu_{i,n}^{1/2} i^{1/2-\nu} n^{-1/2} \mu_{i,n}^{-1/2} \Big) .$$

As in Mason (1991) we bound (1.6) by

$$A\, u^{-1}\, i^{\nu-1/2}\, n^{1/2}\, \mu_{i,n}^{1/2} \exp(-(u^2/8)\, (i^{1-2\nu}/(n\mu_{i,n})))$$

which is the term of a convergent series for any fixed u , which implies

$$P\Big(\max_{1\le i\le n} \sup_{|x-i/n|\le 1/2n} n^{1/2} \frac{|B_n(g_n(x))-B_n(\tilde{g}_n(i/n))|}{i^{1/2-\nu}} \ge u\, \Delta_n \le \lambda \Big)$$

$$\le \varepsilon$$

for large u .

Since (III.2.5) implies also

$$(1.7) \qquad \Big\| \sup_{|s|\le 1/2n} n^{1/2} \frac{|g_n(t+s)-\tilde{g}_n(t)|}{(t(1-t))^{1/2}} \Big\|_n = O(1) \text{ , as } n\to\infty$$

the same technique shows that

$$(1.8) \qquad P\Big(\max_{1\le i\le n} \sup_{|x-i/n|\le 1/2n} n^{1/2} \frac{|B_n(\tilde{g}_n(x))-B_n(\tilde{g}_n(i/n))|}{i^{1/2-\nu}} \ge u \Big) \le \varepsilon$$

for large u. Then (1.7) and (1.8) imply that $\|\delta_{2,n}\|_n = O_p(1)$. Using (1.3) , we conclude that (1.2) is also $O_p(1)$, which is Theorem III.2.1. □

2) <u>Proof of Theorem III.2.2</u>

Introduce the exchangeable process

$$W_n(t) := \Sigma_{i=1}^{[nt]} W_{i,n} \text{ , } 0 \le t \le 1$$

Then we have the representation

(2.1) $$\mathbb{D}_{W,n} \overset{d}{=} \mathbb{W}_n \circ \mathbb{D}_n \circ \zeta_n \quad \text{where}$$

$$\zeta_n := g_n^{\leftarrow} \circ g_{W,n} .$$

Despite the fact that (2.1) holds in distribution, we shall also denote its r.h.s. by $\mathbb{D}_{W,n}$. Using (III.2.11) and linear bounds on $\mathbb{D}_n$, g_n and $g_{W,n}$, we obtain for any $\lambda > 0$,

$$n^{\nu} \sup_{\lambda/n \le \mathbb{D}_n \circ \zeta_n \le 1-\lambda/n} \frac{|n^{1/2}(\mathbb{D}_{W,n} - \mathbb{D}_n \circ \zeta_n) - \sigma B \circ \mathbb{D}_n \circ \zeta_n|}{(I(1-I))^{1/2-\nu}} = O_p(1) .$$

Using linear bounds and Theorem III.2.1, we get

$$a_{W,n} :=$$

(2.2) $$n^{\nu} \left\| \frac{n^{1/2}(\mathbb{D}_{W,n} - \mathbb{D}_n)\ n^{1/2}(g_n - g_n \circ \zeta_n) + B_n \circ \tilde{g}_n - B_n \circ \tilde{g}_n \circ \zeta_n - \sigma\ B \circ \mathbb{D}_n \circ \zeta_n}{(I(1-I))^{1/2-\nu}} \right\|$$

$$= O_p(1) \quad \text{as} \quad n \to \infty .$$

As in Csörgo and Mason (1989) we have

(2.3) $$n^{\nu} \left\| \frac{B \circ \mathbb{D}_n \circ \zeta_n - B \circ g_{W,n}}{(I(1-I))^{1/2-\nu}} \right\|_n = O_p(1) .$$

Since (III.2.14) holds, apply Lemma 3.3 in Barbe (1992b) to obtain

(2.4) $$n^{\nu} \left\| \frac{B \circ g_{W,n} - B \circ \tilde{g}_n}{(I(1-I))^{1/2-\nu}} \right\|_n = O_p(1) .$$

Finally, (2.2)-(2.4) and (III.2.12)-(III.2.13) give (III.2.15). □

3) Proof of Theorem III.3.1

Let $(\omega_i)_{i\ge1}$ be an i.i.d. sequence of exponential r.v.'s with mean 1. Recall that we defined $\Omega_j := \omega_1 + \ldots + \omega_j$ $(j \ge 1)$ to be their partial sums. It is well known (see e.g. Galambos (1978)) that for any fixed k ,

$$\lim_{n\to\infty} P\{ n(1-U_{n-j,n}) \le x_j : 1\le j\le k \} = P\{ \Omega_j \le x_j : 1\le j\le k \} . \tag{3.1}$$

Hence, we need to find some conditions under which

$$p_n := P\{ m_n(\mathbb{U}_n^{\leftarrow}(1-n^{-1}) - \mathbb{U}_{W,n}^{\leftarrow}(1-jc_n n^{-1})) \le x_j : 1\le j\le k \mid \mathbb{U}_n \}$$

converges a.s. or in probability to the r.h.s. of (3.1).

For this, consider the exchangeable process $\mathbb{W}_n(t)$ defined in section 2, and, conditionally on $\mathbb{U}_n$, the following holds $\mathbb{U}_{W,n} \overset{d}{=} \mathbb{W}_n \circ \mathbb{U}_n$ and

$$\mathbb{U}_{W,n}^{\leftarrow} \overset{d}{=} \mathbb{U}_n^{\leftarrow} \circ \mathbb{W}_n^{\leftarrow} . \tag{3.2}$$

With representation (3.2) we obtain

$$p_n = P\{ m_n(\mathbb{U}_n^{\leftarrow}(1-1/n) - \mathbb{U}_n^{\leftarrow}\circ\mathbb{W}_n^{\leftarrow}(1-jc_n/n)) \le x_j : 1\le j\le k \mid \mathbb{U}_n \}$$

$$= P\{ \mathbb{W}_n^{\leftarrow}(1-jc_n/n) \ge \mathbb{U}_n(\mathbb{U}_n^{\leftarrow}(1-1/n) - (x_j/m_n)) : 1\le j\le k \mid \mathbb{U}_n \} .$$

Now, our proof follows that of Deheuvels, Mason and Shorack (1992). From Kiefer (1972) we have

$$1 - \mathbb{U}_n^{\leftarrow}(1-1/n) = O(n^{-1}\text{loglog } n) \quad \text{a.s.}$$

which leads to

$$p_n = P\{ \mathbb{W}_n^{\leftarrow}(1-jc_n/n) \ge \mathbb{U}_n(1-x_j/m_n+O(n^{-1}\text{loglog } n)) : 1\le j\le k \mid \mathbb{U}_n \}$$

almost surely. From Theorem 2, p.604, of Shorack and Wellner (1986) we deduce that for $m_n = o(n/\text{loglog } n)$,

$$|\mathbb{U}_n(1-x_j/m_n+O(n^{-1}\text{loglog } n)) - 1 + x_j/m_n + O(n^{-1}\text{loglog } n)|$$

$$= O((\text{loglog } n)/nm_n)^{1/2} ,$$

so that p_n is equal to

$$P\{\ \mathbb{W}_n^{\leftarrow}(1-jc_n/n) \ge 1 - x_j/m_n + O(((\log\log n)/nm_n)^{1/2} : 1\le j\le k \mid \mathbb{U}_n\ \}$$

$$= P\{\ \mathbb{W}_n^{\leftarrow}(jc_n/n) \le x_j/m_n + O(((\log\log n)/nm_n)^{1/2} : 1\le j\le k \mid \mathbb{U}_n\ \} \text{ a.s.}$$

Therefore, the bootstrap is a.s. consistent if

$$P\{\ m_n\mathbb{W}_n^{\leftarrow}(jc_n/n) \le x_j\ (1+o(1)) : 1\le j\le k \mid \mathbb{U}_n\ \}$$

converges a.s. to the r.h.s. of (3.1), or similarly

$$(3.3) \qquad \{\ m_n\mathbb{W}_n^{\leftarrow}(jc_n/n) \le x_j : 1\le j\le k\ \} \xrightarrow{d} \{\ \Omega_j : 1\le j\le k\ \}\ .$$

The process $N(t) := \#\{\ j : \Omega_j \le t\ \}$ is a Poisson process. Define

$$N_n(t) := \#\{\ j : m_n\mathbb{W}_n^{\leftarrow}(jc_n/n) \le t\ \}\ .$$

Then, (3.3) shows that for any fixed $T > 0$,

$$\{\ N_n(t) : 0 < t < T\ \} \xrightarrow{d} \{\ N(t) : 0 < t < T\ \}\ .$$

But we have the obvious equality

$$N_n(t) = \#\{\ j : j \le n\mathbb{W}_n(t/m_n)/c_n\ \} = [n\mathbb{W}_n(t/m_n)/c_n]$$

which leads to (III.3.7) .

The sufficiency part of the "in probability statement" of Theorem III.3.1 is proved along the same lines.

To obtain the necessary part, observe that for any $c > 0$,

$$p_n = P\{\ n(1-\mathbb{W}_n^{\leftarrow}(1-jc_n/n) \le n(1-\mathbb{U}_n(U_{n-1,n}-cx_j/m_n) : 1\le j\le k \mid \mathbb{U}_n\ \}$$

$$(3.4) \qquad \stackrel{d}{=} P\{\ n\mathbb{W}_n^{\leftarrow}(jc_n/n) \le n\ \mathbb{U}_n(U_{1,n} + cx_j/m_n) : 1 \le j \le k \mid \mathbb{U}_n\ \}\ .$$

Now, recall the fact that

$$\{\ U_{i,n} : 1\le i\le n\ \} \stackrel{d}{=} \{\ \Omega_i/\Omega_{n+1} : 1\le i\le n\ \}\ .$$

Hence, we can rewrite (3.4) as

$$(3.5)\qquad p_n \overset{d}{=} P\{\ n\mathbb{W}_n^{\leftarrow}(jc_n/n) \le N(N^{\leftarrow}(1)(n/\Omega_{n+1})cnx_j/m_n)\ :\ 1\le j\le k\ \mid$$

$$N(n) = n+1\ \}$$

The law of the iterated logarithm ensures that there exists a subsequence $\phi(n)$ such that a.s. for n large

$$\phi(n)/\Omega_{\phi(n)+1} \ge 1 \quad\text{and}\quad \lim_{n\to\infty} \phi(n)/\Omega_{\phi(n)+1} = 1 \quad \text{a.s.}$$

Then, along this subsequence, (3.5) yields

$$p_n \overset{d}{=} P\{\ n\mathbb{W}_n^{\leftarrow}(jc_n/n) \le N(N^{\leftarrow}(1) + cx_jn/m_n)\ :\ 1\le j\le k\ \mid\ N(n) = n+1\ \}$$

for $n = \phi(k)$ and k large enough. Since N is a step function, $x_j \longmapsto N(N^{\leftarrow}(1) + cx_j)$ is also a step function, and therefore, p_n cannot converge to some continuous function of x_j except if $\lim_{n\to\infty} n/m_n = \infty$. Therefore, for the bootstrap to be consistent, we need (III.3.8) □

4) <u>Proof of Theorems III.3.2 and III.3.3</u>

The proof follows exactly the same lines as that of Theorem 2.2 in Deheuvels, Mason and Shorack (1992) : (III.3.17) may be rewritten as

$$(Q(\mathbb{U}_{W,n}^{\leftarrow}(1-jc_n/n) - Q(\mathbb{U}_n^{\leftarrow}(1-1/m_n)))/\hat{\alpha}_n\ .$$

Using Theorem III.3.1 we also have

$$\mathbb{U}_{W,n}^{\leftarrow}(1-jc_n/n) = U_{n,n} - m_n^{-1}\Omega_j + o_p(m_n^{-1})$$

so that (III.3.17) becomes

$$\frac{Q(U_{n,n}-\Omega_j/m_n+o(m_n^{-1})) - Q(U_{n-[n/m_n],n})}{Q(U_{n-[n/m_n],n}) - Q(U_{n-[2n/m_n],n})}$$

$$= \frac{Q\left(1-(m_n(1-U_{n,n})+\Omega_j+o(1))/m_n\right)-Q\left(1-m_n^{-1}m_n(1-U_{n-[n/m_n],n})\right)}{Q\left(1-m_n^{-1}m_n(1-U_{n-[n/m_n],n})\right)-Q\left(1-m_nm_n^{-1}(1-U_{n-[2n/m_n],n})\right)}\ .$$

Since under the assumptions of Theorem III.3.2.i we have

$$\lim_{n\to\infty} 1/m_n = 0\ ,$$

$$\lim_{n\to\infty} m_n(1-U_{n,n}) = 0 \quad \text{in probability,}$$

$$\lim_{n\to\infty} m_n(1-U_{n-[n/m_n]}) = 1 \quad \text{in probability,}$$

then (III.3.16) leads to the result.

The a.s. version is similar. □

Proof of Theorem III.3.3 : Proceed as in Deheuvels, Mason and Shorack (1992) or as in M. Csörgő, S. Csörgő, Horváth and Mason (1986b). □

APPENDIX 1

EXCHANGEABLE VARIABLES OF SUM 1

In this appendix we recall some results which are used along this monograph and concern triangular arrays of row exchangeable r.v.'s. Further results and references may be found, for instance, in Aldous (1983) and Diaconis and Freedman (1987).

Consider $\{ W_{i,n} : 1 \le i \le n , n \ge 1 \}$ a triangular array of rowwise exchangeable r.v.'s such that

(A.1.1) $$W_{i,n} \ge 0 ,$$

(A.1.2) $$\Sigma_{1 \le i \le n} W_{i,n} = 1$$

(A.1.3) $$\limsup_{n \to \infty} E(nW_{1,n} - 1)^4 < \infty$$

(A.1.4) $$n^{1/2} \{ n^{-1} \Sigma_{1 \le i \le n} (nW_{i,n} - 1)^2 - \sigma_{W,n}^2 \} = O_P(1) \quad \text{for some } \sigma_{W,n}^2 > 0 .$$

Define i_n the first random subscript such that $W_{i_n,n} > 0$ and j_n the last (random) subscript such that $W_{j_n,n} > 0$. We shall assume that

(A.1.5) $$\liminf_{\varepsilon \to 0} \liminf_{n \to \infty} P\{ nW_{i_n,n} > \varepsilon \} = 1 .$$

Under these assumptions, Einmahl and Mason (1992) obtained a weighted approximation of the exchangeable process

$$\mathcal{E}_n(t) := n^{1/2} \Sigma_{1 \le i \le [nt]} (W_{i,n} - n^{-1}) .$$

Then, they define the inverse process

$$\mathcal{H}_n(t) := n^{1/2} (t - \mathbb{W}_n^{\leftarrow}(t))$$

where $\mathbb{W}_n^{\leftarrow}$ is the inverse function of $\mathbb{W}_n$,

$$\mathbb{W}_n^{\leftarrow}(s) := \inf\{\ t : \mathcal{H}_n(t) \geq s\ \}\ .$$

Theorem A.1.1 (Einmahl and Mason (1992)) : **Under (A.1.1)-(A.1.5) and on a rich enough probability space there exists probabilistically equivalent versions $\tilde{\mathcal{E}}_n$ and $\tilde{\mathcal{H}}_n$ of $\mathcal{E}_n$ and $\mathcal{H}_n$ and a fixed Brownian bridge B such that for all $0 \leq \nu < 1/4$ and $\tau > 0$,**

(A.1.6) $$\sup_{\tau/n\leq t\leq 1-\tau/n} |\tilde{\mathcal{E}}_n(t) - \sigma_{W,n}B(t)|/(t(1-t))^{(1/2)-\nu} = O_P(n^{-\nu})\ ,$$

(A.1.7) $$\sup_{\tilde{W}_{j_n,n}\leq t\leq \tilde{W}_{j_1,n}} |\tilde{\mathcal{H}}_n(t) - \sigma_{W,n}B(t)|/(t(1-t))^{(1/2)-\nu} = O_P(n^{-\nu})\ .$$

As Einmahl and Mason (1992) show in their Theorem 1, condition A.1.1 is not required for (A.1.6) to hold. They also provide linear bounds for $H_n(.)$ and $H_n^{\leftarrow}(.)$ so that their paper enables us to use the standard techniques of weighted approximation to prove limit theorems (see e.g. the volume edited by Hahn, Mason, Weiner (1991)).

The following result has an important role in chapter I and III in connection with conditions (I.1.1) and (I.3.5). It shows that the covariance structure of exchangeable r.v.'s with sum 1 are subject to some restrictions and that higher moments mimic moments of i.i.d. r.v.'s in some sense.

If $y := (y_1, \ldots, y_n)$ is a vector in $\mathbb{R}^n$ and if $p := (p_1, \ldots, p_n)$ is a multiindex in $\mathbb{N}^n$, we denote $y^p := \Pi_{i=1}^n y_i^{p_i}$. We also define

$$d(p) := \#\{\ i : p_i = 1\ \} \quad \text{and}$$

$$m(p) := \#\{\ i : p_i > 0\ \}\ .$$

$s(p)$ denotes the largest integer smaller than $(d(p)+1)/2$, i.e.

$$s(p) := [(d(p)+1)/2]\ .$$

Finally, we define $|p| := p_1 + \ldots + p_n$.

Proposition A.1.2 : **Let $W(.) = (W(1), \dots, W(n))$ be an exchangeable random vector, such that**

$$\Sigma_{i=1}^{n} W(i) = 0 \ . \tag{A.1.8}$$

Let $p = (p_1, \dots, p_n)$ be a multiindex. Then

$$|E_W W(.)^p| \le m(p)^{s(p)} E_W |W(1)|^{|p|+1} / (n-m(p)+s(p))^{s(p)} \ . \tag{A.1.9}$$

Proof of proposition A.1.2 : Using exchangeability of the $W(i)$'s, in the left hand side of (A.1.9), we may assume that $p_1, \dots, p_r \ge 2$, $p_{r+1} = \dots = p_m = 1$ and $p_{m+1} = \dots = p_n = 0$ with $m = m(p)$ and $r = r(p) = m(p) - d(p)$.

From (A.1.8) we deduce

$$E_W W(1)^{p_1} \dots W(m-1)^{p_{m-1}} \Sigma_{i=1}^{n} W(i) = 0 \ . \tag{A.1.10}$$

Then, exchangeability of the $W(i)$'s implies that the l.h.s. of (A.1.10) is

$$\Sigma_{\delta_1+\dots+\delta_{m-1}=1} E_W W(1)^{p_1+\delta_1} \dots W(m-1)^{p_{m-1}+\delta_{m-1}}$$

$$+ (n-m+1) E_W W(1)^{p_1} \dots W(m)^{p_m} = 0 \ ,$$

where the δ_i 's are 0 or 1. Consequently

$$\begin{aligned} &|E_W W(.)^p| \\ &\le (m-1)(n-m+1)^{-1} \max_{\delta_1+\dots+\delta_{m-1}=1} |E_W W(1)^{p_1+\delta_1} \dots W(m-1)^{p_{m-1}+\delta_{m-1}}| \\ &= \frac{(m-1)}{(n-m+1)} \max_{\delta_1+\dots+\delta_{m-1}=1} |E_W W(1)^{p_1+\delta_1} \dots W(r+1)^{p_{r+1}+\delta_{r+1}} \\ &\qquad W(r+2) \dots W(m-1)| \ . \end{aligned} \tag{A.1.11}$$

This inequality may be iterated q times to get

$$|E_W W(.)^p| \le \frac{(m-1)\dots(m-q)}{(n-m+1)\dots(n-m+q))} \max_{\delta_1+\dots+\delta_{r+q}=q} |E_W W(1)^{p_1+\delta_1} \dots$$

$$W(r+q)^{p_{r+q}+\delta_{r+q}} W(r+q+1) \dots W(m-q)| \ ,$$

and we stop when $m - q = r + q$ or $m - q = r + q + 1$.

If $m - q = r + q$ we get the bound

(A.1.12) $$|E_W W(.)^p|$$

$$\le A(n,m,q) \max_{\delta_1+..+\delta_{r+q}=1} |E_W W(1)^{p_1+\delta_1} \dots W(r+q)^{p_{r+q}+\delta_{r+q}}|$$

with

(A.1.13) $$A(n,m,q) := \frac{(m-1)\dots(m-q)}{(n-m+1)\dots(n-m+q)} \ ,$$

and if $m - q = r + q + 1$, we apply (A.1.11) one more time to obtain

$$|E_W W(.)^p| \le A(n,m,q+1)$$

$$\max_{\delta_1+..+\delta_{r+q+1}=1} |E_W W(1)^{p_1+\delta_1} \dots W(r+q+1)^{p_{r+q+1}+\delta_{r+q+1}}| \ .$$

Then, Hölder's inequality shows that for any multiindex $(s_1 , \dots, s_k)$ with $s_i > 0$ ($1 \le i \le k$),

$$|E_W W(1)^{s_1} \dots W(k)^{s_k}| \le E|W_W(1)|^{|s|} \ ,$$

and therefore, (A.1.12) and (A.1.13) yield

(A.1.14) $$|E_W W(.)|^p \le A(n,m,m-q-r) \ E_W|W(1)|^{|p|+1} \ .$$

Since $q = [(m-r)/2]$, we have $m-q-r = [(m-r+1)/2] = [(d+1)/2]$ and (A.1.14) implies (A.1.9). □

As mentioned in chapter II, an easy way to generate exchangeable r.v. with sum equal to 1 is first to generate a sample $Y_{1,n} , \dots, Y_{n,n}$ of i.i.d. r.v.'s with common distribution function $\mathcal{Y}_n$ and to define

(A.1.15) $$W_{i,n} := Y_{i,n}/(Y_{1,n} + \dots + Y_{n,n}) .$$

To compare the moments of the $W_{i,n}$ and that of the $Y_{i,n}$ we assume that there exists $\alpha , \beta , \gamma > 0$ such that for all $n > 1$,

(A.1.16) $$\mathcal{Y}_n(x) \le \alpha x^{\beta} \text{ if } 0 \le x \le \gamma , \text{ and}$$

(A.1.17) $$\limsup_{n\to\infty} E\, Y_{i,n}^8 < \infty .$$

Notice that if (A.1.16) holds for some $\beta = \beta_1$, then it holds for any $\beta \le \beta_1$.

Proposition A.1.3 : **For the weights defined in (A.1.15), under (A.1.16) and (A.1.17), we have**

$$\sigma_{W,n}^2 = \sigma_{Y,n}^2 + n^{-1}(3\sigma_{Y,n}^4 - \sigma_{Y,n}^2 - 2\sigma_{Y,n}^3\beta_{Y,n}) + O(n^{-3/2}) ,$$

$$\beta_{W,n} = \beta_{Y,n} + O(n^{-1/2}) \quad \textbf{as} \quad n\to\infty .$$

Proof of Proposition A.1.2 : There is no loss of generality to assume that $E\, Y_{i,n} = 1$. Define $\overline{Y}_n = n^{-1}\Sigma_{i=1}^n Y_{i,n}$. For the variance, notice that

$$E(Y_{i,n}^2/\overline{Y}_n^2) = E\, Y_{i,n}^2(1 - 2(\overline{Y}_n-1) + 3(\overline{Y}_n-1)^2) - 4\, E(Y_{i,n}^2(\overline{Y}_n-1)^3\overline{Y}_n^{-2}$$

$$- 3\, E(Y_{i,n}^2(\overline{Y}_n-1)^4\overline{Y}_n^{-2} .$$

An easy calculation shows that

$$E\, Y_{i,n}^2(1 - 2(\overline{Y}_n-1) + 3(\overline{Y}_n-1)^2)$$

$$= \sigma_{Y,n}^2 + 1 + n^{-1}(3\sigma_{Y,n}^4 - \sigma_{Y,n}^2 - 2\sigma_{Y,n}^3\beta_{Y,n}) + O(n^{-3/2}) .$$

Then, Hölder's inequality yields

$$|E(Y_{i,n}^2(\overline{Y}_n-1)^3\overline{Y}_n^{-2}| \le (E\, Y_{i,n}^6)^{1/3}\, (E(\overline{Y}_n-1)^6)^{1/3}\, (E\, \overline{Y}_n^{-4})^{1/2} .$$

Marcinkiewicz-Zygmund's theorem implies under (A.1.17)

$$E(\overline{Y}_n - 1)^6 = O(n^{-3/2}) \quad \text{as} \quad n \to \infty .$$

Then, under (A.1.17), we have for any $x \le \gamma$

$$\mathcal{Y}_n * \mathcal{Y}_n(x) = \int \mathcal{Y}_n(x-t) d\mathcal{Y}_n(t) \le \alpha^2 \beta \, x^{\beta} \, \Gamma(\beta+1)\Gamma(\beta)/\Gamma(2\beta+1)$$

and using an induction, for $n \ge 5$

$$E \, \overline{Y}_n^{-4} \mathbb{I}\{ \overline{Y}_n \le \delta \} = \int_0^{\delta} x^{-4} d\mathcal{Y}^{*n}(x) \le c^n$$

where c is a constant that may be taken smaller than 1 by taking β small in (A.1.16) and this ends the proof of the result on the variance.

To prove the result on the kurtosis, proceed in the same way, noticing that

$$E(Y_{i,n}/\overline{Y}_n)^3 = E \, Y_{i,n}^3 + E(Y_{i,n}(\overline{Y}_n - 1)/\overline{Y}_n)^3$$

$$= E \, Y_{i,n}^3 + O(n^{-1/2}) \quad \text{as} \quad n \to \infty . \quad \square$$

APPENDIX 2

EXPRESSION OF $E_W Q_{W,N}{}^P$ AS A SUM OF U-STATISTICS

In this appendix we express $E_W Q_{W,n}{}^p$ as a sum of U-statistics and give conditions under which

(A.2.1) $$E_P | E_W \, Q_{W,n}{}^p |^r < \infty$$

From expression (IV.4.3) we obtain

$$Q_{W,n} = Q_{W,n,1} + 2^{-1} Q_{W,n,2} + Q_{W,n,3} \quad \text{with}$$

$$Q_{W,n,1} := n^{-1/2} \Sigma_{i=1}^n W_n(i) \, T_1(i) \, ,$$

$$Q_{W,n,2} := n^{-3/2} \Sigma_{i=1}^n \Sigma_{j=1}^n W_n(i) \, W_n(j) \, T_2(i,j) \, ,$$

$$Q_{W,n,3} := n^{-1/2} \Sigma_{i=1}^n W_n(i) \, (\, n^{-1} \Sigma_{j=1}^n T_2(i,j) \,) \, .$$

Therefore, we can express $E_W Q_{W,n}{}^p$ as a sum of terms

$$q(p_1,p_2,p_3) := E_W \, Q_{W,n,1}{}^{p_1} \, Q_{W,n,2}{}^{p_2} \, Q_{W,n,3}{}^{p_3}$$

with $p_1 + p_2 + p_3 = p$. Consequently, all what we need to show is that $q(p_1,p_2,p_3)$ may be decomposed as a sum of U-statistics and to give conditions under which

$$E_P | q(p_1,p_2,p_3) |^r < \infty \, .$$

Now, denote

$$s := s(p_1,p_2,p_3) = p_1 + 2p_2 + 2p_3$$

and define

$$c(i_1,\ldots,i_s) := E_W W_n(i_1)\ldots W_n(i_{s-p_3}) ,$$

$$t(i_1,\ldots,i_s) := T_1(i_1) \ldots T_1(i_{p_1}) T_2(i_{p_1+1}, i_{p_1+p_2+1}) \ldots$$

$$T_2(i_{p_1+p_2}, i_{p_1+2p_2}) T_2(i_{p_1+2p_2+1}, i_{p_1+2p_2+p_3+1}) \ldots$$

$$T_2(i_{p_1+2p_2+p_3}, i_{p_1+2p_2+2p_3}) .$$

So, we can rewrite

$$\text{(A.2.2)} \qquad q(p_1,p_2,p_3) = \Sigma^n_{i_1=1} \ldots \Sigma^n_{i_s=1} c(i_1,\ldots,i_s)\ t(i_1,\ldots,i_s) .$$

Therefore, we have the decomposition

$$q(p_1,p_2,p_3) = q'(p_1,p_2,p_3) + \Sigma^{s-1}_{k=1} q_k(p_1,p_2,p_3) \quad \text{with}$$

$$q'(p_1,p_2,p_3) := \Sigma^n_{i_1=1} \ldots \Sigma^n_{i_{s-1}=1} \Sigma^n_{\begin{cases} i_s=1 \\ i_s \notin \{i_1,\ldots,i_{s-1}\} \end{cases}} c(i_1,\ldots,i_s)$$

$$t(i_1,\ldots,i_s)$$

$$q_k(p_1,p_2,p_3) := \Sigma^n_{i_1=1} .. \Sigma^n_{i_{s-1}=1} c(i_1,..,i_{s-1},i_k)\ t(i_1,..,i_{s-1},i_k).$$

Now, assume that we have a statistics of the form

$$\tau := \Sigma^n_{i_1=1} \ldots \Sigma^n_{i_k=1} \Sigma_{\begin{cases} 1 \le i_{k+1} \le n \\ i_{k+1} \notin \{i_1,\ldots,i_k\} \end{cases}} \Sigma^n_{\begin{cases} i_s=1 \\ i_s \notin \{i_1,\ldots,i_{s-1}\} \end{cases}}$$

$$\gamma(i_1,\ldots,i_s)\ \tau(i_1,\ldots,i_s)$$

as are $q'(p_1,p_2,p_3)$ and all the $q_k(p_1,p_2,p_3)$ for all $1 \le k \le s$ (for these last statistics, we have to change, in τ , s by $s-1$ and define $\gamma(i_1,\ldots,i_{s-1}) = \gamma(i_1,\ldots,i_{s-1},i_k)$ and similarly for $\tau(i_1,\ldots,i_{s-1},i_k)$ and $t(i_1,\ldots,i_{s-1},i_k)$). Then, τ may be rewritten as

$$\tau = \tau' + \Sigma^{k-1}_{j=1} \tau_j' \quad \text{with}$$

$$\tau' := \Sigma^n_{i_1=1} \cdots \Sigma^n_{i_{k-1}=1} \Sigma_{\begin{cases} 1\le i_k \le n \\ i_k \notin \{i_1,\dots,i_k\} \end{cases}} \Sigma^n_{\begin{cases} i_s=1 \\ i_s \notin \{i_1,\dots,i_{s-1}\} \end{cases}}$$

$$\gamma(i_1,\dots,i_s)\ \tau(i_1,\dots,i_s) \quad \text{and}$$

$$\tau_j' := \Sigma^n_{i_1=1} \cdots \Sigma^n_{i_{k-1}=1} \Sigma_{\begin{cases} 1\le i_{k+1} \le n \\ i_{k+1} \notin \{i_1,\dots,i_{k-1}\} \end{cases}} \Sigma^n_{\begin{cases} i_s=1 \\ i_s \notin \{i_1,\dots,i_{s-1}\} \end{cases}}$$

$$\gamma(i_1,\dots i_{k-1}, i_j, i_{k+1},\dots,i_s)\ \tau(i_1,\dots i_{k-1}, i_j, i_{k+1},\dots,i_s)\ .$$

Clearly, up to some changes of notations, τ' and the τ_j' are of the same form as τ , and we can apply this decomposition recursively, until we obtain terms of the form

$$\tilde{\tau}_k := \Sigma^n_{i_1=1} \cdots \Sigma_{\begin{cases} 1\le i_2 \le n \\ i_2 \notin \{i_1\} \end{cases}} \cdots \Sigma^n_{\begin{cases} i_k=1 \\ i_k \notin \{i_1,\dots,i_{k-1}\} \end{cases}}$$

$$\gamma_k(i_1,\dots,i_k)\ \tau_k(i_1,\dots,i_k)$$

(an example of such a decomposition is given in Appendix A.3).

Applying this decomposition scheme to $q(p_1,p_2,p_3)$ as written in (A.2.2), we can express it as a sum of terms

(A.2.3)
$$\Sigma^n_{i_1=1} \cdots \Sigma_{\begin{cases} 1\le i_2 \le n \\ i_2 \notin \{i_1\} \end{cases}} \cdots \Sigma^n_{\begin{cases} i_k=1 \\ i_k \notin \{i_1,\dots,i_{k-1}\} \end{cases}}$$
$$c(i_{\sigma(1)},\dots,i_{\sigma(k)}, i_{m_1},\dots,i_{m_{s-k}})$$

$$t(i_{\sigma(1)},\dots,i_{\sigma(k)}, i_{m_1},\dots,i_{m_{s-k}})$$

where $m_1 ,\dots, m_{s-k}$ belong to $\{ 1 , \dots, k \}$ and σ is a permutation of k elements.

Now, due to the exchangeability of the $W_{i,n}$'s, the coefficients

$$c(i_{\sigma(1)},\dots,i_{\sigma(k)}, i_{m_1},\dots,i_{m_{s-k}})$$

in (A.2.3) depend only on $(\sigma(1),\ldots,\sigma(k),m_1,\ldots,m_{s-k})$, so that (A.2.3) may be rewritten as

$$\tilde{c}(\sigma(1),\ldots,\sigma(k),m_1,\ldots,m_{s-k})\ U(\sigma(1),\ldots,\sigma(k),m_1,\ldots,m_{s-k})$$

$$U(\sigma(1),\ldots,\sigma(k),m_1,\ldots,m_{s-k})$$

$$:= \Sigma^n_{i_1=1}\ldots\Sigma^n_{\begin{cases} i_k=1 \\ i_k\notin\{i_1,\ldots,i_{k-1}\}\end{cases}} t(i_{\sigma(1)},\ldots,i_{\sigma(k)},i_{m_1},\ldots,i_{m_{s-k}}) ,$$

which is a U-statistic by a last symmetrization. Hence, we can rewrite $E_W Q_{W,n}{}^p$ as a sum of U-statistics.

Now, to ensure (A.2.1), we have to find conditions under which we have

$$\text{(A.2.4)} \qquad E_P|U(\sigma(1),\ldots,\sigma(k),m_1,\ldots,m_{s-k})|^r < \infty$$

for all possible indexes $(\sigma(1),\ldots,\sigma(k),m_1,\ldots,m_{s-k})$. From Lemma 5.2.2.A in Serfling (1980), we deduce that (A.2.4) holds if

$$\text{(A.2.5)} \qquad E_P|t(i_{\sigma(1)},\ldots,i_{\sigma(k)},i_{m_1},\ldots,i_{m_{s-k}})| < \infty .$$

Then, Proposition I.2.1 implies that (A.2.5) holds if

$$E_P|H(X)|^{2pr} < \infty . \quad \square$$

APPENDIX 3

DERIVATION OF $E_W(Q_{W,N}^2)$ AND S_N^2

Using the differentiability of T at P , we have

$$S_n^2 = E_W(n^{1/2}(T(P_{W,n})-T(P)) - n^{1/2}(T(P_n)-T(P)))^2$$

$$= E_W\ S_{W,n}^2 + o_p(n^{-1/2}) ,$$

where $S_{W,n} := S_{W,n,1} + S_{W,n,2} + S_{W,n,3}$ and with

$$S_{W,n,i} := n^{1/2}(1/i!)\ (D^iT(P_{W,n}-P)^i - D^iT(P_n-P)^i) .$$

More explicitly,

$$S_{W,n,1} = n^{-1/2}\ \Sigma_{i=1}^n\ (W_i^{(n)}-1)\ T_1(i) ,$$

$$S_{W,n,2} = \frac{1}{2n^{3/2}}\ \Sigma_{i=1}^n\Sigma_{j=1}^n(W_i^{(n)}W_j^{(n)}-1)\ T_2(i,j) , \text{ and}$$

$$S_{W,n,3} = \frac{1}{6n^{5/2}}\ \Sigma_{i=1}^n\Sigma_{j=1}^n\Sigma_{k=1}^n(W_i^{(n)}W_j^{(n)}W_k^{(n)}-1)\ T_3(i,j,k) .$$

Notice that $S_{W,n,i} = O_{P,W}(n^{-(i-1)/2})$.

We now evaluate each term in the sum obtained in developing $E_W(S_{W,n,1} + S_{W,n,2} + S_{W,n,3})^2$. We limit ourself to the only terms that matter in approximating S_n^2 up to $o_p(n^{-1})$. The first term is

$$E_W S_{W,n,1}^2 = E_W W_n(1)^2\ n^{-1}\Sigma_{i=1}^n T_1(i)^2$$

$$+ E_W W_n(1) W_n(2)\ n^{-1}\Sigma_{i=1}^n \Sigma_{\substack{j=1 \\ j\neq i}}^n\ T_1(i)\ T_1(j) .$$

In order to evaluate $E_W S_{W,n,2}^2$,we denote [i,j,k,l] the sum of terms of the form $\omega(i,j,k,l) := E_W(W_i^{(n)}W_j^{(n)}-1)(W_k^{(n)}W_l^{(n)}-1)T_2(i,j)T_2(k,l)$ when i, j, k, l run over 1, ..., n . We denote [i,j',k,l] the same sum but with $j\neq i$. More generally, we put prime (') at an indice to denote the sum with this indice different from those which appear before it in the bracket. Hence, [i,j',k',l] denotes the sum of $\omega(i,j,k,l)$ over $j\neq i$, $k\neq i$ and $k\neq j$ while [i,j',k',j'] is the sum of $\omega(i,j,k,j)$ over $j\neq i$, $k\neq i$ and $k\neq j$. Now,

$$4\, E_W S^2_{W,n,2} = [i,j,k,l]$$

$$= [i,j',k,l] + [i,i,k,l]$$

(A.3.1)
$$= [i,j',k',l'] + [i,j',k',i,] + [i,j',k',j'] + [i,j',k',k']$$

$$+ [i,j',i,l'] + [i,j',i,i] + [i,j',i,j']$$

$$+ [i,j',j',l'] + [i,j',j',i] + [i,j',j',j']$$

$$+ [i,i,k',l'] + [i,i,k',i] + [i,i,k',k'] + [i,i,i,l'] + [i,i,i,i].$$

Notice the symmetries [i,j,k,l] = [j,i,l,k] = [k,l,i,j] or similarly

$$[i,i,i,l'] = \Sigma^n_{i=1}\Sigma^n_{\substack{l=1\\ l\neq i}} E_W(W_i^{(n)2}-1)(W_i^{(n)}W_l^{(n)}-1)T_2(i,i)T_2(i,l)$$

$$= [i,j',i,i] \text{ , and}$$

$$[i,j',k',j'] = [i,j',k',i] = [i,j',i,l'] \text{ ,}$$

which lead to

(A.3.2)
$$4\, E_W S^2_{W,n,2} = [i,j',k',l'] + 4[i,j',k',i] + 2[i,j',k',k']$$

$$+ 4[i,j',i,i] + 2[i,j',i,j'] + [i,i,k',k'] + [i,i,i,i] \text{ .}$$

We now evaluate the expectations

$$c(i,j,k,l) := E_W(W_i^{(n)}W_j^{(n)}-1)(W_k^{(n)}W_l^{(n)}-1)$$

$$= E\,(W_n(i)W_n(j) + W_n(i) + W_n(j))\,(W_n(k)W_n(l) + W_n(k) + W_n(l))$$

that we need. Using Appendix 1, we have

$$c(1,2,3,4) = c(1,2,3,3) + O(n^{-1}) = O(n^{-1}) ,$$

$$c(1,2,3,1) = E_W W_n(1)^2 + O(n^{-1}) ,$$

$$c(1,2,1,1) = E_W W_n(1)^3 + 2\, E_W W_n(1)^2 + O(n^{-1}) ,$$

$$c(1,2,1,2) = E_W W_n(1)^2 W_n(2)^2 + 2\, E_W W_n(1)^2 + O(n^{-1}) ,$$

$$c(1,1,2,2) = E_W W_n(1)^2 W_n(2)^2 + O(n^{-1}) ,$$

$$c(1,1,1,1) = E_W W_n(1)^4 + 4\, E_W W_n(1)^3 + 4\, E_W W_n(1)^2 .$$

Consequently, noticing also that all the brackets with primes in (A3.2) are U-statistics (eventually up to a symetrization), we have

$$4\, E_W\, S^2_{W,n,2} = 4\, E_W W_n(1)\, n^{-3} \Sigma^n_{i=1} \Sigma^n_{\substack{j=1\\ j\neq i}} \Sigma^n_{\substack{k=1\\ k\neq i,j}}\; T_2(i,j)\; T_2(k,i)$$

$$+ 4\, (\, E_W W_n(1)^3 + 2E_W W_n(1)^2\,)\; n^{-3}\; \Sigma^n_{i=1} \Sigma^n_{\substack{j=1\\ j\neq i}}\; T_2(i,j)\; T_2(i,i)$$

$$+ 2\, (E_W W_n(1)^2 W_n(2)^2 + 2E_W W_n(1)^2\,)\; n^{-3}\; \Sigma^n_{i=1} \Sigma^n_{\substack{j=1\\ j\neq i}}\; T_2(i,j)^2$$

$$+ E_W W_n(1)^2 W_n(2)^2\; n^{-3}\; \Sigma^n_{i=1} \Sigma^n_{\substack{j=1\\ j\neq i,j}}\; T_2(i,i)\; T_2(j,j) + o_p(n^{-1})$$

$$= 4\, E_W W_n(1)^2\; n^{-2}\; \Sigma^n_{i=1} \Sigma^n_{\substack{j=1\\ j\neq i,j}}\; E\; T_2(X,i)\; T_2(X,j)$$

$$+ 2\, (\, E_W W_n(1)^2 W_n(2)^2 + 2E_W W_n(1)^2\,)\; n^{-1}\; E\; T_2(X,Y)^2$$

$$+ E_W W_n(1)^2 W_n(2)^2\; n^{-1}\; (E\; T_2(X,X))^2 + o_p(n^{-1}) .$$

We now handle $E_W\, S_{W,n,1} S_{W,n,3}$. Using a similar convention as before for $E_W\, S^2_{W,n,2}$, we agree to denote $[i,j,k,l]$ the sum of terms of the form $E_W\, (W^{(n)}_i - 1)\; (W^{(n)}_j W^{(n)}_k W^{(n)}_k - 1)\; T_1(i)\; T_3(j,k,l)$. We make the same convention with the superscript prime over a letter in a bracket. Then (A.3.1) still holds, and using symmetries of the bracket with primed letters,

$$6\, E_W\, S_{W,n,1} S_{W,n,3} = [i,j',k',l'] + 3[i,i,j',k'] + 3[i',j',j',k']$$

$$+\ 2[i,i,i,j'] + 3[i,i,j',j'] + 2[i,j',j',j'] + [i,i,i,i]\ .$$

We now evaluate the expectations

$$c(i,j,k,l) := E_W (W_i^{(n)} - 1)\ (W_j^{(n)} W_k^{(n)} W_k^{(n)} - 1)$$

$$= E_W W_n(i)\ (\ W_n(j) W_n(k) W_n(l) + W_n(j) W_n(k) + W_n(j) W_n(l)$$

$$+\ W_n(k) W_n(l) + W_n(j) + W_n(k) + W_n(l)\ )$$

that we need. We have

$$c(1,2,3,4) = c(1,2,2,3) + O(n^{-1}) = c(1,2,2,2) + O(n^{-1}) = O(n^{-1})\ ,$$

$$c(1,1,2,3) = E_W W_n(1)^2 + O(n^{-1})\ ,$$

$$c(1,1,1,2) = E_W W_n(1)^3 + 2\, E_W W_n(1)^2 + O(n^{-1})\ ,$$

$$c(1,1,2,2) = E_W W_n(1)^2 W_n(2)^2 + E_W W_n(1)^2 + O(n^{-1})\ ,$$

$$c(1,1,1,1) = E_W W_n(1)^4 + 3\, E_W W_n(1)^3 + 3\, E_W W_n(1)^2\ .$$

Consequently,

$$6\, E_W\, S_{W,n,1} S_{W,n,3} = 3\, E_W W_n(1)^3\, n^{-3} \sum_{i=1}^n \sum_{\substack{j=1\\ j\neq i}}^n \sum_{\substack{k=1\\ k\neq i,j}}^n T_1(i)\, T_3(i,j,k)$$

$$+\ 2\ (\ E_W W_n(1)^3 + 2\, E_W W_n(1)^2\ )\ n^{-3} \sum_{i=1}^n \sum_{\substack{j=1\\ j\neq i}}^n T_1(i)\, T_3(i,i,j)$$

$$+\ 3\ (\ E_W W_n(1)^2 W_n(2)^2 + E_W W_n(1)^2\ )\ n^{-3} \sum_{i=1}^n \sum_{\substack{j=1\\ j\neq i}}^n T_1(i)\, T_3(i,j,j)$$

$$+\ o_p(n^{-1})$$

$$= 3\ E_W W_n(1)^3\ n^{-2}\ \Sigma_{i=1}^n \Sigma_{\substack{j=1\\ j\neq i}}^n\ E\ T_1(X)\ T_3(X,i,j)$$

$$+\ 3\ (\ E_W W_n(1)^2 W_n(2)^2 + E_W W_n(1)^2\ )\ n^{-1}\ E\ T_1(X)\ T_3(X,Y,Y)$$

$$+\ o_p(n^{-1})\ .$$

We finaly need the following term, where now

$$c(i,j,k) := E_W(W_i^{(n)}-1)(W_j^{(n)}W_k^{(n)}-1)\ ,$$

$$2\ E_W\ S_{W,n,1}S_{W,n,2} = c(1,2,3)\ n^{-2}\ \Sigma_{i=1}^n \Sigma_{\substack{j=1\\ j\neq i}}^n \Sigma_{\substack{k=1\\ k\neq i,j}}^n\ T_1(i)\ T_2(j,k)$$

$$+\ 2\ c(1,2,1)\ n^{-2}\ \Sigma_{i=1}^n \Sigma_{\substack{j=1\\ j\neq i}}^n T_1(i)\ T_2(i,j)$$

$$+\ c(1,2,2)\ n^{-2}\ \Sigma_{i=1}^n \Sigma_{\substack{j=1\\ j\neq i}}^n T_1(i)\ T_2(j,j)$$

$$+\ c(1,1,1)\ n^{-2}\ \Sigma_{i=1}^n T_1(i)\ T_2(i,i)\ .$$

We calculate $c(i,j,k) = E_W\ W_n(i)\ (\ W_n(j)W_n(k) + W_n(j) + W_n(k)\ )$ for the indices that we need, namely,

$$c(1,2,3) = c(1,2,2) + O(n^{-1}) = O(n^{-1})\ ,$$

$$c(1,2,1) = E_W W_n(1)^2 + O(n^{-1})\ ,$$

$$c(1,1,1) = E_W W_n(1)^3 + 2\ E_W W_n(1)^2\ .$$

Consequently,

$$2\ E_W\ S_{W,n,1}S_{W,n,2} = 2\ E_W W_n(1)^2\ n^{-2}\ \Sigma_{i=1}^n \Sigma_{\substack{j=1\\ j\neq i}}^n T_1(i)\ T_2(i,j)$$

$$+\ (E_W W_n(1)^3 + 2\ E_W W_n(1)^2\ )\ n^{-1}\ E_P T^{(1)}(X,P)\ T^{(2)}(X,X,P) + o_p(n^{-1})\ .$$

Putting all the pieces together yields (V.4.2), assuming (as assumed in section V.4) that

$$E_W(nW_{1,n}-1)^2(nW_{2,n}-1)^2 = E_W W_n(1)^2 W_n(2)^2 = (E_W W_n(1))^2 \ (1+o(1)) \ .$$

the expectation of $E_W \ Q_{W,n}^2$ is obtained in the same way, but truncating all the terms above at the order $o_p(n^{-1/2})$.

APPENDIX 4

CALCULATION OF $E_W Q^3_{W,N}$

From definition (IV.4.3) we have

$$\text{(A.4.1)} \qquad \sigma^3_{W,n} Q^3_{W,n} = \left(n^{-1/2} \Sigma^n_{i=1} W_n(i)T_1(i)\right)^3$$

$$+ 3\left(n^{-1/2}\Sigma^n_{i=1}W_n(i)T_1(i)\right)^2 2^{-1}n^{-3/2}\Sigma^n_{i=1}\Sigma^n_{j=1}(W^{(n)}_i W^{(n)}_j - 1)T_2(i,j)$$

$$+ 3\left(n^{-1/2}\Sigma^n_{i=1}W_n(i)T_1(i)\right)\left(2^{-1}n^{-3/2}\Sigma^n_{i=1}\Sigma^n_{j=1}(W^{(n)}_i W^{(n)}_j -1)T_2(i,j)\right)^2$$

$$+ \left(2^{-1} n^{-3/2} \Sigma^n_{i=1}\Sigma^n_{j=1}(W^{(n)}_i W^{(n)}_j - 1)T_2(i,j)\right)^3 .$$

Now we calculate the expectation of each of the four terms in the r.h.s. of (A.4.1) following the procedure described in Appendix 2.

First, let

$$A_1 = E_W\left(n^{-1/2} \Sigma^n_{i=1} W_n(i)T_1(i)\right)^3$$

$$\text{(A.4.2)} \qquad =n^{-3/2} E_W \Sigma^n_{i=1}\Sigma^n_{j=1}\Sigma^n_{k=1} W_n(i)W_n(j)W_n(k)\ T_1(i)T_1(j)T_1(k)$$

When we have a sum over i,j,k as in (A.3.4) we can denote it by $[i,j,k]$. To denote the summation $\Sigma^n_{i=1}\Sigma^n_{\substack{j=1\\ j\neq i}}\Sigma^n_{k=1}$ we just use $[i,j',k]$, the $'$ added to the j meaning that the summation is taken for j different of all preceeding indexes. Similarly $\Sigma^n_{i=1}\Sigma^n_{\substack{j=1\\ j\neq i}}\Sigma^n_{\substack{k=1\\ k\neq i\\ k\neq j}}$ is denoted $[i,j',k']$.

Now we have

$$A_1 = [i,j,k] = [i,j',k] + [i,i,k]$$

$$= [i,j',k'] + [i,j',i] + [i,j',j] + [i,i,k'] + [i,i,i]$$

$$= E_W W_n(1)W_n(2)W_n(3)\ U_1 + E_W W_n(1)^2 W_n(2)\ U_2 + E_W W_n(1)W_n(2)^2\ U_3$$

$$+ E_W W_n(1)^2 W_n(2)\ U_4 + E_W W_n(1)^3\ U_5$$

Where $U_1, \ldots, U_5$ are U-statistics defined by :

$$U_1 := n^{-3/2} \Sigma_{i=1}^n \Sigma_{\substack{j=1\\ j\neq i}}^n \Sigma_{\substack{k=1\\ k\neq i\\ k\neq j}}^n T_1(i)T_1(j)T_1(k)\ ,$$

$$U_2 := U_3 = U_4 = n^{-3/2} \Sigma_{i=1}^n \Sigma_{\substack{j=1\\ j\neq i}}^n T_1(i)T_1(j)^2\ , \text{ and}$$

$$U_5 := n^{-3/2} \Sigma_{i=1}^n T_1(i)^3\ .$$

Notice that U_2 is really a U-statistic since it may be rewritten as

$$U_2 = n^{-3/2} \Sigma_{i=1}^n \Sigma_{\substack{j=1\\ j\neq i}}^n 2^{-1}\left(T_1(i)T_1(j)^2 + T_1(j)T_1(i)^2\right)\ .$$

Next, if $E_P|H(X)|^3 < \infty$, Lemma IV.4.1 and Lemma 5.2.2.A in Serfling (1980) show that

$$E|A_1 - E_W W_n(1)^3\ U_5| = O(n^{-1}) \quad \text{as} \quad n\to\infty\ .$$

Therefore , using Markov inequality

$$A_1 = E_W W_n(1)^3 U_5 + O_P(n^{-1}) = E_W W_n(1)^3\ n^{-1/2}\ E_P T_1(i)^3 + O_P(n^{-1})\ .$$

Then, we deal with

$$A_2 := (3/2)\ E_W\ n^{-5/2} \Sigma_{i=1}^n \Sigma_{j=1}^n \Sigma_{k=1}^n \Sigma_{l=1}^n W_n(i)W_n(j)(W_k^{(n)} W_l^{(n)} - 1)$$

$$T_1(i)T_1(j)T_1(k,l)\ ,$$

which is the second term in (A.4.1). Using the same kind of notation as in the formal equality (A.3.1) and symmetries, we obtain

(A.4.3) $\quad A2 = [i,j',k',l'] + 4[i,j',k',i] + [i,j',k',k']$

$$+ 2[i,j',i,i] + 2[i,j',i,j']$$

$$+ [i,i,k',l'] + [i,i,k',i] + [i,i,k',k']$$

$$+ [i,i,i,l'] + [i,i,i,i] .$$

Now we have to calculate the various expectations

$$c(i,j,k,l) = E_W W_n(i) W_n(j) \left(W_k^{(n)} W_l^{(n)} - 1 \right)$$

$$= E_W W_n(i) W_n(j) \left(W_n(k) W_n(l) + W_n(k) + W_n(l) \right)$$

according to the respective values of i, j, k, l.

$$c(1,2,3,4) = O(n^{-2}) ,$$

$$c(1,2,3,1) = c(1,2,3,3) + O(n^{-1}) = c(1,2,1,1) + O(n^{-1})$$

$$= c(1,1,2,3) + O(n^{-1}) = c(1,1,2,1) + O(n^{-1}) = c(1,1,1,2) + O(n^{-1})$$

$$= O(n^{-1})$$

$$c(1,2,1,2) = c(1,2,2,1) + O(n^{-1}) = c(1,1,2,2) + O(n^{-1})$$

$$= E_W W_n(1)^2 W_n(2)^2 + O(n^{-1}) ,$$

$$\gamma(1,1,1,1) = E_W W_n(1)^4 - 2\, E_W W_n(1)^3 .$$

Hence, with (A.4.3), we deduce that

$$A_2 = E_W W_n(1)^2 W_n(2)^2\, 2\, n^{-5/2} \Sigma_{i=1}^n \Sigma_{\substack{j=1 \\ j \neq i}}^n T_1(i) T_1(j) T_2(i,j)$$

$$+ E_W W_n(1)^2 W_n(2)^2\, n^{-5/2} \Sigma_{i=1}^n \Sigma_{\substack{j=1 \\ j \neq i}}^n T_1(i)^2\, T_2(j,j) + A_2'$$

where $E_P |A_2'| = O(n^{-1})$.

Then, from lemma 5.2.2.A in Serfling (1980), we deduce

$$A_2 = n^{-1/2} E_W W_n(1)^2 W_n(2)^2 E(2T_1(i)T_1(j)T_2(i,j) + T_1(i)^2 T_2(j,j)) + A_2''$$

$$= n^{-1/2} E_W W_n(1)^2 W_n(2)^2 E(2T_1(i)T_1(j)T_2(i,j) + S^2 T_2(i,i)) + A_2''$$

where $A_2'' = O_P(n^{-1})$ and $E_P|nA_2''|^{1+\delta} = O(n^{-1})$ as $n \to \infty$,

Now let us examin the third term in the r.h.s. of (A.3.1), namely

$$A_3 := n^{-7/2} E_W \Sigma_{i=1}^n \Sigma_{j=1}^n \Sigma_{k=1}^n \Sigma_{l=1}^n \Sigma_{m=1}^n W_n(i)(W_j^{(n)} W_k^{(n)} - 1)(W_l^{(n)} W_m^{(n)} - 1)$$

$$T_1(i)T_1(j,k)\ T_2(l,m)\ .$$

Then using the same kind of notations as before, the formal equality (A.3.1) yields

(A.4.4) $$A3 = [i,j',k',l',m] + [i,j',k',i,m] + [i,j',k',j,m]$$

$$+ [i,j',k',k,m] + [i,j',i,l',m] + [i,j',i,i,m] + [i,j',i,j,m]$$

$$+ [i,j',j,l',m] + [i,j',j,i,m] + [i,j',j,j,m]$$

$$+ [i,i,k',l',m] + [i,i,k',i,m] + [i,i,k',k,m]$$

$$+ [i,i,i,l',m] + [i,i,i,i,m]\ .$$

Then, with

$$c(i,j,k,l,m) := E_W W_n(i)(W_j^{(n)} W_k^{(n)} - 1)(W_l^{(n)} W_m^{(n)} - 1)$$

$$= E_W W_n(i)\{W_n(j)W_n(k) - W_n(j) - W_n(k)\}\{W_n(l)W_n(m) - W_n(l) - W_n(m)\}$$

we have

$$c(1,2,3,4,5) = O(n^{-2})\ ,$$

$$c(1,2,3,4,1) = O(n^{-1})\ ,$$

$$c(1,2,3,4,2) = c(1,2,3,4,3) = O(n^{-1})\ ,$$

$$c(1,2,3,4,4) = O(n^{-1})\ ,$$

so that

$$[i,j',k',l',m] = O_p(n^{-1}) \quad \text{and}$$

(A.4.5) $$E_P\big|[i,j',k',l',m]\big|^{1+\delta} = O_p(n^{-1-\delta}) \quad \text{as} \quad n\to\infty .$$

Then it is easy, using the same technique, to see that each term in (A.4.4) satisfies a property like (A.4.5) provided that

$$\limsup_{n\to\infty} E_W W_n(1)^5 < \infty \quad \text{and}$$

$$E_P H(X)^{5+\delta} < \infty .$$

It remains to deal with the fourth term in the r.h.s. of (A.4.1),

$$A_4 := n^{-9/2} E_W \Sigma_{i=1}^n \Sigma_{j=1}^n \Sigma_{k=1}^n \Sigma_{l=1}^n \Sigma_{m=1}^n \Sigma_{p=1}^n (W_i^{(n)} W_j^{(n)} - 1)(W_k^{(n)} W_l^{(n)} - 1)$$

$$(W_m^{(n)} W_p^{(n)} - 1)\; T_2(i,j)\; T_2(k,l)\; T_2(m,p) .$$

If the following holds,

$$\limsup_{n\to\infty} E_W\; W_n(1)^6 < \infty \quad \text{and}$$

$$E_P H(X)^{6+\delta} < \infty ,$$

one can check that

$$E_P|A_4|^{1+\delta} = O(n^{-1-\delta}) \quad \text{as} \quad n\to\infty .$$

Hence, we can rewrite (A.4.1) as

$$E_W\, Q^3_{W,n} = E_W W_n(1)^3\, n^{-1/2}\, E_P T_1(1)^3$$

$$+ (3/2)\, n^{-1/2}\, E_W W_n(1)^2 W_n(2)^2\, E_P\big(2T_1(1)T_1(2)T_2(1,2) + S^2\, T_2(1,1)\big)$$

$$+ \mathrm{Rem}_n$$

with

$$E_P |Rem_n|^{1+\delta} = O(n^{-1-\delta}) \quad \text{as} \quad n \to \infty ,$$

so that Markov's inequality implies

$$P\{ n^{1/2} |Rem_n| > \varepsilon_n \} \leq \varepsilon_n^{-1} n^{-1/2-\delta/2} = o(n^{-1/2})$$

for a suitably chosen sequence $(\varepsilon_n)_{n \geq 0}$ converging to 0 (e.g. $\varepsilon_n = 1/\log n$).

APPENDIX 5

FINITE SAMPLE ASYMPTOTIC FOR THE MEAN AND THE BOOTSTRAP OF THE MEAN

An extremely simple way to get more informations on how the weights have to be chosen is to look at finite sample properties. Of course, it is extremely difficult to investigate the behaviour of the generalized bootstrapped distribution of the mean for a finite sample size. However, we can easily study the tail behaviour, or equivalently, some large deviations for a finite sample size. A quite complete theory for these large deviations for the mean has been recently built by Broniatowski (1992).

Here we deal with a very elementary case as an heuristic. Indeed, the aim of this appendix is only to make a remark which suggests how to restrict the possible choices of the weights. In the following, we assume that the X_i's have a Pareto tail

(A.5.1) $$P\{ X_i \geq x \} = x^{-\rho}\ell_P(x) \quad \text{as } x\to\infty$$

where $\ell_P(.)$ is a slowly varying function at infinity, which means that for any $\lambda > 0$,

(A.5.2) $$\lim_{x\to\infty} \ell_P(\lambda x)/\ell_P(x) = 1$$

(see e.g. Bingham, Goldie and Teugels (1987)). We assume that the weights $Y_{i,n}$ are i.i.d. and nonnegative with

(A.5.3) $$1 - \mathcal{Y}_n(x) = x^{-\alpha}\ell_{\mathcal{Y}}(x) .$$

Let $\overline{X}_n := n^{-1}\Sigma_{i=1}^n X_i$ be the empirical mean and $\overline{X}_{Y,n} = n^{-1}\Sigma_{i=1}^n Y_{i,n} X_i$ its generalized bootstrapped version. Then the following holds and is an easy consequence of section XIII.5 in Feller (1971).

Theorem A.5.1 : **Assume that (2.2) and (2.4) hold with** $\operatorname{supp} \mathcal{Y}_n \subset [0,\infty)$ **, then**

(A.5.4) $\lim_{x\to\infty} P\{ \overline{X}_n > x \} / P\{ \overline{X}_{Y,n} > x \mid P_n \} \in (0,\infty)$ P-a.s.

if and only if

(A.5.5) $\alpha = \rho$ **and** $\ell_{\mathcal{Y}}(x) \sim \ell_P(x)$ **as** $x\to\infty$.

Moreover, if (A.5.5) holds, then

(A.5.6) $\lim_{x\to\infty} P\{ \overline{X}_n > x \} / P\{ \overline{X}_{Y,n} > \zeta_n x \mid P_n \} = 1$ P-a.s.

with $\zeta_n := (n^{-1} \Sigma_{i=1}^n X_i^{\alpha} \mathbb{I}\{X_i > 0\})^{1/\alpha}$.

Hence, from (A.5.5) we deduce that the tail of the weights should mimic that of the sample. Unfortunately this tail is unknown. Two ways are possible : either we estimate the tail index and use the estimator to define $\mathcal{Y}_n$, either we consider that, for practical purpose, instead of a limit as $x\to\infty$ we just have a "large" x , and in this case we are not interested in the very upper tail. In the second case, a possible practical solution is just to use the empirical distribution function for $\mathcal{Y}_n$, maybe after having smoothed it. Some simulation results (available on request) shows that the second method gives better results.

APPENDIX 6

WEIGHTS GIVING AN ALMOST SURELY CONSTISTENT BOOTSTRAPPED MEAN

The aim of this appendix is mainly to show that some results of chapter II.4 are not always compatible with other reasonable requirements and to exhibit a strange behaviour of the bootstrap.

Consider again the case of the mean $\overline{X}_n$ and its bootstrapped version $\overline{X}_{W,n}$. One can require $\overline{X}_{W,n}$ to be a consistent estimator. Such question has been investigated in the case of Efron's bootstrap by Athreya (1983), Athreya, Ghosh, Low and Sen (1984), S. Csörgő (1991), Arenal, Cuesta-Albertos and Matran (1994) (who mention some errors in Csörgő's (1991) proofs).

We denote by $\mathcal{W}_n$ the distribution function of the n-th row of the triangular array of the weights $\{ W_{i,n} : 1 \le i \le n , n \ge 1 \}$. W_n may depend on the sample $X_1 , \dots, X_n$. To be more precise, we consider the sequence $\{X_i : i\ge1\}$ defined on a probability space $(\Omega, \mathcal{A}, P)$. Then, let $\{U_{i,n} : 1\le i\le n; n\ge1\}$ be a sequence of i.i.d. $\mathcal{U}_{[0,1]}$ r.v.'s on some probability space $(\Omega^*, \mathcal{A}^*, P^*)$, independent of the sequence $(X_i)_{i\ge1}$. The weights $\{ W_{i,n} : 1\le i\le n \}$ at the step n are supposed to be generated with $\{ U_{i,n} : 1\le i\le n \}$ using Skorokhod's (1976) construction. By this, we emphazise that when W_n is chosen, the weights are independent of the data and given the sequence $(\mathcal{G}_n)_{n\ge1}$, the rows $\{ W_{i,n} : 1 \le i \le n \}$ at different steps n are independent. Then we can look under which conditions $\overline{X}_{W,n}$ converges P^*-a.s. P-a.s. to E_PX . By the Borel-Cantelli lemma, this is equivalent to find conditions under which, for any $\varepsilon > 0$

(A.6.1) $$\Sigma_{n\ge1}P^*\{ |\overline{X}_{W,n} - E_PX| > \varepsilon \} < \infty \quad \text{P-a.s.}$$

The results of this section are consequence of those of Barbe (1992a) where the proofs are given.

We shall assume that, for some $\theta > 1$,

(A.6.2) $$E_P|X_1|^\theta < \infty .$$

Theorem A.6.1 : **If (A.6.2) holds and if**

(A.6.3) $$\Sigma_{n\geq 1} n^{\eta}\, E_W(|W_n(1)|^{2r} \mid X_1,\dots,X_n) < \infty \quad \text{P-a.s.}$$

with

$$\eta(r,\theta) := \begin{cases} 2r(\theta^{-1}-1) & \text{if } 1 < \theta < 2 \\ -r+\delta & \text{if } \theta \geq 2\ , \text{ with } \delta > 0 \text{ arbitrary small.} \end{cases}$$

then, (A.6.1) holds.

The condition (A.6.3) is in fact almost a necessary condition if the weights are multinomial (Efron's bootstrap) or i.i.d. For Efron's bootstrap, we have :

Theorem A.6.2 : **If** $\{\, m_n W_{i,n} : 1 \leq i \leq n \,\}$ **has a multinomial** $\mathcal{M}(m_n,1/n,\dots,1/n)$ **distribution and** $(m_n)_{n\geq 1}$ **is a monotone sequence, then** (A.6.1) **implies**

$$\Sigma_{n\geq 1}\ 1 - F(2m_n) + F(-2m_n) < \infty\ .$$

For the bootstrap with i.i.d. weights, we assume that

(A.6.4) $\{\, nW_{i,n} : 1 \leq i \leq n\ ,\ n \geq 1 \,\}$ are i.i.d. with common d.f. $\mathcal{Y}$ which is supposed to be independent of n .

and define

$$\mathcal{Y}^*(x) := P\{\, n|W_{1,n} - W_{2,n}| \leq x \,\} = \int (\mathcal{Y}(y+x)-\mathcal{Y}(y-x))\, d\mathcal{Y}(y)\ .$$

We shall assume that

(A.6.5) $1 - \mathcal{Y}^*$ is regularly varying with index $-\beta$ at esssup H

which means that for any $\lambda > 0$,

$$\lim_{x\to \text{esssup } H} (1-\mathcal{Y}^*(\lambda x))/(1-\mathcal{Y}^*(x)) = \lambda^{-\beta}$$

(see e.g. Bingham, Goldie and Teugels (1987)).

Theorem A.6.3 : **If** (A.6.1), (A.6.4)-(A.6.5) **hold, then**

(A.6.6) $$\Sigma_{n\geq 1}\ 1 - F(4n/\mathcal{Y}^{*\leftarrow}(1-n^{-1})) + F(-4n/\mathcal{Y}^{*\leftarrow}(1-n^{-1})) < \infty\ .$$

Condition (A.6.6) is an Hölder type condition . Indeed if $1-F$ is regularly varying with index $-\theta$ as $x\to\infty$, (A.6.6) asserts that $E|nW_{1,n}|^{\beta}$ is finite for any β such that $\beta^{-1}+\theta^{-1}\leq 1$, while (A.6.3) gives the same condition.

The interesting fact is that in the situation of Theorem A.6.3, $\beta^{-1}+\theta^{-1}\leq 1$ is a necessary and sufficient condition for the bootstrapped mean to be a.s. consistent. If $1<\theta<2$, this condition is not compatible with condition (A.5.5). Therefore, we cannot simultaneously estimate correctly the tail of the distribution of the mean for a finite sample size and require the bootstrapped mean to be a.s. consistent when the sample has an infinite variance!

APPENDIX 7

DIFFERENTIAL OF THE FUNCTIONAL $T^{(1)}(x,P)$

Let us first obtain the differentials of $T^{(1)}(x,P)$, $T^{(2)}(x,y,P)$, as a function of P.

Since

$$T^{(1)}(x,Q) = (d/dt)\ T(\ (1-t)Q+t\delta_x)\ )|_{t=0}$$

we obtain

$$T^{(1)}(x,Q)^{(1)}(y,P) = (\partial^2/\partial t\ \partial t')\ T(\ (1-t')((1-t)P+t\delta_y)) + t'\delta_x)$$

Expanding $T(\ (1-t')((1-t)P+t\delta_y)) + t'\delta_x)$ as a function of $T^{(1)}(x,P)$ and $T^{(2)}(x,y,P)$ and taking the cross-derivatives, we obtain

(A.7.1) $$T^{(1)}(x,P)^{(1)}(y,P) = T^{(2)}(x,y,P) - T^{(1)}(y,P) .$$

Similarly, we have

$$T^{(2)}(x,y,P)^{(1)}(z,P)$$

$$= (\partial^3/\partial t\ \partial t'\ \partial t'')\ T(\ (1-t-t')((1-t'')P+t''\delta_z) + t\delta_x + t'\delta_y\ )$$

$$= T^{(3)}(x,y,z,P) - T^{(2)}(x,z,P) - T^{(2)}(y,z,P)$$

and finally

$$T^{(1)}(x,P)^{(2)}(y,z,P) = T^{(3)}(x,y,z,P) - 2\ T^{(2)}(y,x,P) .$$

Now, let

$$f(Q) := (T(Q)-T(P))/S(Q) = (T(Q)-T(P))/S^2(Q)^{1/2} .$$

The gradient of f(.) is given by

(A.7.2) $$f^{(1)}(x,Q) = T^{(1)}(x,Q)/S(Q) - (1/2)(T(Q)-T(P))S^{(1)}(x,Q)/S(Q)^3 ,$$

which yields

$$f^{(1)}(x,P) = T^{(1)}(x,P)/S(P) .$$

Using (A.7.1), we deduce from the derivation of (A.7.2) that

$$f^{(2)}(x,y,P) = T^{(2)}(x,y,P)/S(P)$$
$$- (1/2)(T^{(1)}(x,P)S^{(1)}(y,P) + T^{(1)}(y,P)S^{(1)}(x,P))/S(P)^3 \quad \text{and}$$

$$\begin{aligned} f^{(3)}(x,y,z,P) = {} & T^{(3)}(x,y,z,P)/S(P) \\ & - (1/2)\left(T^{(2)}(x,y,P)S^{(1)}(z,P) + T^{(2)}(x,z,P)S^{(1)}(y,P) \right. \\ & \left. + T^{(2)}(y,z,P)S^{(1)}(x,P)\right)/S(P)^3 \\ & - (1/2)\left(T^{(1)}(x,P)S^{(2)}(y,z,P) + T^{(1)}(y,P)S^{(2)}(x,z,P) \right. \\ & \left. + T^{(1)}(z,P)S^{(2)}(x,y,P)\right)/S(P)^3 \\ & + (3/4)\left(T^{(1)}(x,P)S^{(1)}(y,P)S^{(1)}(z,P) \right. \\ & + T^{(1)}(y,P)S^{(1)}(x,P)S^{(1)}(z,P) \\ & \left. + T^{(1)}(z,P)S^{(1)}(y,P)S^{(1)}(x,P)\right)/S(P)^5 . \end{aligned}$$

APPENDIX 8

EDGEWORTH EXPANSIONS UP TO $o(N^{-1})$ FOR A THIRD ORDER DIFFERENTIABLE STATISTICAL FUNCTIONAL AND ITS STUDENTIZED FORM

Let $\mathcal{S}(B^k,\mathbb{R})$ be the set of the functions from B^k to $\mathbb{R}$ symmetric in their arguments. Let further $\mathcal{S}$ be the set of all applications Δ from $\{1,..,k\}$ to $\{1,..,k\}$, $i(\Delta):=\{\Delta(1),\Delta(2),\dots,\Delta(k)\}$ is the image of $\{1,..,k\}$ by Δ.
We define

$$L_p^{(k)}(P) := \{\ g \in \mathcal{S}(B^k,\mathbb{R}), \qquad \int |g(x_{\Delta(1)},..x_{\Delta(k)})|^p \prod_{j\in i(\Delta)} P(dx_j) < \infty\ ,$$

$$\Delta \in S\}$$

Notice that this definition allows repetitions of a given x_i in $x_{\Delta(1)},\dots,x_{\Delta(k)}$ so that we integrate $g(x_{\Delta(1)},\dots,x_{\Delta(k)})$ over all the different x_j, $j\in i(\Delta)$. The space $L_p^{(k)}(P)$ is endowed with the norm

$$\| g \|_{p,P}^{(k)} = \Big(\sum_{\Delta\in\mathcal{S}} \int..\int |g(x_{\Delta(1,k)})|^p \prod_{j\in i(\Delta)} P(dx_j) \Big)^{1/p}$$

is a Banach space (see Filippova (1961)).

Let $\psi_n(t)$ be the characteristic function of $n^{1/2}(T(P_n) - T(P)) / S(P)$. We shall assume that we have

$$\text{(A.8.1)} \qquad \int_{n^{2/3}}^{n\log n} |\psi_n(t)|\ dt = o(n^{-1}) \quad \text{(see another condition in A.8.11)}$$

Condition (A.8.1) is actually implied by the condition (A.8.11) below which is somehow easier to check (see after the proposition (A.8.1) the discussion of the hypotheses). We need also the following strong Cramér condition
for any $t \in \mathbb{R}$,

(A.8.2) $\left|E_F \exp\left(itT^{(1)}(X,F)\right)\right| < 1 - \chi_1(t)$, where $\chi_1(t)$ is a non negative continuous function such that $\lim_{t\to\infty} \chi_1(t) \geq \xi > 0$

Next, we assume that for some $\varepsilon_i > 0$, i=1,..,3

(A.8.3) $T^{(1)}(.,P) \in L^{(1)}_{4+\varepsilon_1}(P)$, **for some** $\varepsilon_1 > 0$,

(A.8.4) $T^{(2)}(.,P) \in L^{(2)}_{3+\varepsilon_2}(P)$, **for some** $\varepsilon_2 > 0$,

(A.8.5) $T^{(3)}(.,P) \in L^{(2)}_{3+\varepsilon_3}(P)$, **for some** $\varepsilon_3 > 0$

Consider now

$$H_n(x,P) = P\{ n^{1/2}(T(P_n) - T(P)) / S(P) \leq x \}$$

In order to state an Edgeworth expansion for $H_n(x,P)$, recall the coefficient $K_{i,j}(P)$ defined among chapters I and II

(A.8.6) $$S(P)\, K_{1,1}(P) = E_P T^{(2)}(X_1,X_1) ,$$

$$S(P)^3\, K_{1,3}(P) = E_P T^{(1)}(X_1,P)^3$$

(A.8.7) $$+ 3E_P T^{(2)}(X_1,X_2,P)T^{(1)}(X_1,P)T^{(1)}(X_2,P) ,$$

(A.8.8) $$S(P)^2\, K_{2,2}(P) = (E_P T^{(2)}(X_1,X_1,P))^2/4 + E_P T^{(2)}(X_1,X_2,P)^2/2$$

$$+ E_P\left(T^{(1)}(X_1,P)\, T^{(2)}(X_1,X_1,P)\right) + E_P\left(T^{(1)}(X_1,P)\, T^{(3)}(X_1,X_2,X_2,P)\right) ,$$

(A.8.9) $$S(P)^4\, K_{2,4}(P) = 2\, S(P)^4\, K_{1,1}(P)\, K_{1,3}(P) + ET^{(1)}(X_1,P)^4 - 3\, S(P)^4$$

$$+ 12\, E_P\left(T^{(1)}(X_1,P)\, T^{(1)}(X_2,P)^2\, T^{(1)}(X_1,X_2,P)\right)$$

$$+ 12\, E_P\left(T^{(1)}(X_2,P)\, T^{(1)}(X_3,P)\, T^{(2)}(X_1,X_2,P)\, T^{(2)}(X_1,X_3,P)\right)$$

$$+ 4\, E_P\left(T^{(1)}(X_1,P)\, T^{(1)}(X_2,P)\, T^{(1)}(X_3,P)\, T^{(3)}(X_1,X_2,X_3,P)\right) ,$$

(A.8.10) $$K_{2,6}(P) = K_{1,3}(P)^2 .$$

Proposition A.8.1 : **Assume that (II.4.4), (II.4.5), (A.8.1), (A.8.5) hold. Then $H_n(x,P)$ has the following Edgeworth expansion**

$$H_n(x,P) = \mathcal{E}_{2,H}(x,P) + o(n^{-1}) \text{ , with}$$

$$\mathcal{E}_{2,H}(x,P) = \Phi(x) - n^{-1/2}\left(\frac{K_{1,1}(P)}{2} + \frac{K_{1,3}(P)}{6}\left(x^2-1\right)\right)\phi(x)$$

$$- n^{-1}\left(\frac{K_{2,2}(P)}{2}x + \frac{K_{2,4}(P)}{24}\left(x^3-3x\right) + \frac{K_{2,6}(P)}{72}\left(x^5 - 10x^3 + 15x\right)\right)\phi(x)$$

a) Discussion of the hypotheses

By the regularity conditions (II.4.4) and (II.4.5) we mean that we may control the remainder of the third order taylor expansion of the functional $T(P_n)$ at P in probability. These conditions may be replaced by the stronger but more rigorous, Gâteaux type condition (See Filippova (1961) and Reeds (1976) for some discussion)

$$P\left(n^{-3/2} \sup_{t\in[0,1]} \left|\frac{d^4T\left((1-t)P_n+tP\right)}{dt^4}\right| > \varepsilon\right) = o(n^{-1}) \text{ , for any } \varepsilon>0 .$$

If T is fourth order differentiable for a metric $d_{\mathcal{H}}$, then it is easy to check this condition, with the tools of chapter I.

The main problem to obtain the Edgeworth expansion up to $o(n^{-1})$ is to control the behaviour of the characteristic function for large values of the argument (see Callaert, Janssen and Veraverbeke (1980), Götze (1984)). The technical condition (A.8.1) which allows that control is implied by the following conditional Cramér condition:

(A.8.11) There exists a non negative constant γ two sequences $p = p_n$ and $q = q_n$ and a borel set A_n with $P(A_n) = o(n^{-1})$, such that, for any $t \in [n^{2/3}, n\log n]$

$$\left| E_P\left(\exp\left(it\, n^{-3/2}\sum_{i=q}^{n}\sum_{j=p}^{n}\left(T^{(1)}(X_1,P) + T^{(1)}(X_i,P) + T^{(1)}(X_j,P)\right)/3 + \left(T^{(2)}(X_1,X_i,P) + T^{(2)}(X_1,X_j,P) + T^{(2)}(X_i,X_j,P)\right)/2 + T^{(3)}(X_1,X_i,X_j,P)/6\right) \,\middle|\, (X_p,..,X_q,...X_n)\in A_n\right)\right| \le e^{-\gamma} .$$

(see a similar smoothness condition in Götze (1984)). Callaert, Janssen and Veraverbeke (1980) in the case of statistics of degree 2, show that such a condition is satisfied, if the distribution of the kernel of the U-statistic $U(X_i,X_j) = \left(T^{(1)}(X_i,P) + T^{(1)}(X_j,P)\right) / 2 + T^{(2)}(X_i,X_j,P)$ admits an absolutely continuous part. Conditions (A.8.3)-(A.8.5) are convenient but may be clearly refined which is not our purpose here.

b) Proof of Proposition A.8.1

To obtain an Edgeworth expansion up to $o(n^{-1})$, we cannot directly apply the results of Bickel, Götze and van Zwet (1986). However, we follow their proof. There is no loss of generality in assuming that $S^2(P) = 1$. Until the end, we write

$$\eta_{n-p}(t) := E\left(\Pi_{j=1}^{n-p} \exp\left(i\frac{t}{\sqrt{n}} T^{(1)}(X_j,P)\right)\right) .$$

The characteristic function $\Psi_n(t)$ is given by

$$\Psi_n(t) := E_P\left(\Pi_{1,n}(t)\, \Pi_{2,n}(n^{-1/2} t/2)\, \Pi_{3,n}(n^{-1} t/6)\right)$$

with

$$\Pi_{1,n}(t) := \exp\left(it \int T^{(1)}(x_1,P)\, \beta_n(dx_1)\right)$$

$$\Pi_{2,n}(t) := \exp\left(it \int T^{(2)}(x_1,x_2,P)\, \beta_n(dx_1)\beta_n(dx_2)\right)$$

$$\Pi_{3,n}(t) := \exp\left(it \int T^{(3)}(x_1,x_2,x_3,P)\, \beta_n(dx_1)\beta_n(dx_2)\beta_n(dx_3)\right)\Bigg)$$

$$\beta_n = n^{1/2}(P_n - P)$$

The following lemma gives the expansion of $\Psi_n(t)$ for central values of t and allows one to bound the remainder uniformly in t. The proof of lemma A.8.2 is postponed to the proof of the theorem.

Lemma A.8.2 : **Under (A.8.2)-(A.8.5), there exists some non negative constants** a , b **depending on** ε_1 , ε_2 , ε_3 **such that for every** $t \in [\,0,\, C\, n^{1/2}]$, **where** C **is a non negative constant, the characteristic function** $\Psi_n(t)$ **admits the expansion**

$$\Psi_n(t) = \Psi_n^E(t) + r_{3,n}(t) \quad \textbf{with}$$

$$\Psi_n^E(t) := \exp(-t^2/2)\left(1 + n^{-1/2}\left(it\,\frac{K_{1,1}(P)}{2} - it^3\frac{K_{1,3}(P)}{6}\right) + n^{-1}\left(-\frac{K_{2,2}(P)}{2}t^2 + \frac{K_{2,4}(P)}{24}t^4 - \frac{K_{2,6}(P)}{72}t^6\right)\right)$$

where the $K_{i,j}(P)$ **are given in** (A.8.6)-(A.8.10) **and**

$$r_{3,n}(t) = t^{1+a}\ o\left(n^{-(1+a)}\right) + t^{(2+b)}\ o\left(n^{-(2+b)/2)}\right) + t^{3+a+b}o\left(n^{-(2+a+b/2)}\right) + t^2\ o(n^{-3/2}) + t^3\ o(n^{-2}).$$

uniformly in $t \in [\ 0,\ C\ n^{1/2}]$.

The function $\Psi_n^E(.)$ is the Fourier transform of $\mathcal{E}_{2,H}(.,P)$. From Esseen's smoothing lemma (see for example Feller (1971) p. 538) we have

$$|H_n(x,P) - \mathcal{E}_{2,H}(.,P)| \le \frac{2}{\pi}\int_0^{n\log n} t^{-1}\ |\Psi_n(t) - \Psi_n^E(t)|\ dt + O((n\log n)^{-1})$$

Put $\lambda := \min(a/2(1+a),\ b/3(2+b),\ 1/9)$, first using the bound of lemma A.8.2 we have

$$\int_0^{n^\lambda} t^{-1}\ |\Psi_n(t) - \Psi_n^E(t)|\ dt = \int_0^{n^\lambda} t^{-1}|r_{3,n}(t)|\ = o(n^{-1})\ .$$

Moreover, we have

$$\int_{n^\lambda}^{n\log n} t^{-1}\ |\Psi_n(t) - \Psi_n^E(t)|\ dt \le \int_{n^\lambda}^{n\log n} t^{-1}\ |\Psi_n(t)|dt + \int_{n^\lambda}^{n\log n} t^{-1}|\Psi_n^E(t)|dt$$

It is clear that $\int_{n^\lambda}^{n\log n} t^{-1}|\Psi_n^E(t)|dt = o(n^{-1})$. Under (A.8.1) we just have to check that $\int_{n^{\lambda}}^{n^{2/3}} |\psi_n(t)|\, dt = o(n^{-1})$ holds

The proof for this is similar to Callaert, Janssen and Veraverbeke (1980) (see lemma 4) or Bickel, Götze and Van Zwet (1986) (see from (2.15) to (2.22) of their proof), who use a conditioning argument, and is omitted. The fact that (A.8.11) implies (A.8.1) may be proved as in Callaert, Janssen and Veraverbeke's (1980) Lemma 6.

c) Proof of Lemma A.8.2

i) Expansion of the characteristic function

Expanding $\Pi_{3,n}(.)$ at 0 , gives, for any $|a| \le 1$,

$$\Pi_{3,n}(n^{-1}t/6) := 1 + it/6\, n^{-1}\int T^{(3)}(x_1,x_2,x_3,P)\Pi_{j=1}^{3}\beta_n(dx_j) + R_{3,n}(t)$$

with

$$R_{3,n}(t) := \left(\left|t_{3,n} \int T^{(3)}(x_1,x_2,x_3,P)\Pi_{j=1}^{3}\beta_n(dx_j)\right|\right)^{1+a}$$

for some $t_{3,n} \in [0,n^{-1}\, t/6]$.

Expanding $\Pi_{2,n}(.)$ at 0 , gives, for any $|b| \le 1$

$$\Pi_{2,n}(n^{-1/2}t/2) := 1 + i\; t/2\; n^{-1/2} \int T^{(2)}(x_1,x_2,P)\; \Pi_{j=1}^{2}\beta_n(dx_j)$$
$$- t^2/8\; n^{-1}\left(\int T^{(2)}(x_1,x_2,P)\; \Pi_{j=1}^{2}\beta_n(dx_j)\right)^2 + R_{2,n}(t)\ ,$$

with

$$R_{2,n}(t) := \left(\left|t_{2,n} \int T^{(2)}(x_1,x_2,P)\; \beta_n(dx_1)\beta_n(dx_1)\right|\right)^{2+b}\ ,$$

for some $t_{2,n} \in [0,n^{-1/2}\; t/2]$.

Thus, we have

$$\Psi_n(t) := E_P\left[\Pi_{1,n}(t)\left(1+ i\; t/2\; n^{-1/2} \int T^{(2)}(x_1,x_2,P)\; \beta_n(dx_1)\beta_n(dx_2)\right)\right.$$

$$- t^2/8\ n^{-1}\Big(\Big(\int T^{(2)}(x_1,x_2,P)\ \beta_n(dx_1)\beta_n(dx_2)\Big)\Big)^2\Big)$$

$$\Big(1 + it/6\ n^{-1}\int T^{(3)}(x_1,x_2,x_3,P)\ \beta_n(dx_1)\beta_n(dx_2)\beta_n(dx_3)\Big)\Big)\Big)$$

$$+ r_{1,n}(t)$$

with

$$r_{1,n}(t) := E_P\big(\Pi_{1,n}(t)\ \ \Pi_{2,n}(n^{-1/2}\ t/2)\ \ R_{3,n}(t)\big)$$

$$+ E_P\big(\Pi_{1,n}(t)\ \ R_{2,n}(t)\ \Pi_{3,n}(n^{-1}t/6)\ \big)$$

$$- E_P\big(\Pi_{1,n}(t)\ \ R_{2,n}(t)\ R_{3,n}(t)\big).$$

It follows that

(A.8.13) $$\Psi_n(t) = S_{1,n}(t) + i\ t/2\ n^{-1/2}\ S_{2,n}(t) - t^2/8\ n^{-1}\ S_{3,n}(t)$$

$$+ i\ t/6\ n^{-1}\ S_{4,n}(t) + r_{2,n}(t)\ ,$$

with

$$S_{1,n}(t) := E_P\big(\Pi_{1,n}(t)\big)\ ,$$

$$S_{2,n}(t) := E_P\Big(\Pi_{1,n}(t)\int T^{(2)}(x_1,x_2,P)\ \Pi_{j=1}^2\beta_n(dx_j)\Big)\ ,$$

$$S_{3,n}(t) := E_P\Big(\Pi_{1,n}(t)\Big(\int T^{(2)}(x_1,x_2,P)\ \Pi_{j=1}^2\beta_n(dx_j)\Big)^2\Big)\ ,$$

$$S_{4,n}(t) := E_P\Big(\Pi_{1,n}(t)\ \int T^{(3)}(x_1,x_2,x_3,P)\ \Pi_{j=1}^3\beta_n(dx_j)\Big)\ ,$$

and

$$r_{2,n}(t) := r_{1,n}(t) - t^2 n^{-3/2}/12\ E_P\Big(\Pi_{1,n}(t)\int T^{(2)}(x_1,x_2,P)\Pi_{j=1}^2\beta_n(dx_j)$$

$$\int T^{(3)}(x_1,x_2,x_3,P)\ \Pi_{j=1}^3\beta_n(dx_j)\Big) - it^3/48\ n^{-2}E_P\Big(\Pi_{1,n}(t)$$

$$\Big(\int T^{(2)}(x_1,x_2,P)\ \Pi_{j=1}^2\beta_n(dx_j)\Big)^2\int T^{(3)}(x_1,x_2,x_3,P)\ \Pi_{j=1}^3\beta_n(dx_j)\Big)\ .$$

ii) Computation of the terms in the expansion

Now use the independence argument. Put $\xi := \frac{t}{\sqrt{n}}$ and $A_n^p := n!/(n-p)!$ then we have

$$S_{1,n}(t) = \eta_n(t) ,$$

(A.8.14)
$$S_{2,n}(t) = n^{-1} A_n^2 E_P\Big(T^{(2)}(X_1,X_2,P) \Pi_{j=1}^2 \exp\big(i\xi\ T^{(1)}(X_j,P)\big) \Big)\eta_{n-2}(t)$$

$$+ n^{-1} A_n^1 E_P\Big(T^{(2)}(X_1,X_1,P) \exp\big(i\xi\ T^{(1)}(X_1,P)\big)\Big)\eta_{n-1}(t) ,$$

(A.8.15)
$$S_{3,n}(t) := n^{-2}\Bigg(A_n^4 E_P\Big(T^{(2)}(X_1,X_2,P)\ T^{(2)}(X_3,X_4,P)$$

$$\Pi_{j=1}^4 \exp\big(i\xi\ T^{(1)}(X_j,P)\big)\Big)\eta_{n-4}(t)$$

$$+ 2\ A_n^3 E_P\Big(T^{(2)}(X_1,X_1,P)T^{(2)}(X_3,X_2,P)\Pi_{j=1}^3 \exp\big(i\xi\ T^{(1)}(X_j,P)\big)\Big)\eta_{n-3}(t)$$

$$+ 4\ A_n^3 E_P\Big(T^{(2)}(X_1,X_2,P)T^{(2)}(X_1,X_3,P)\Pi_{j=1}^3 \exp\big(i\xi\ T^{(1)}(X_j,P)\big)\Big)\eta_{n-3}(t)$$

$$+ A_n^2 E_P\Big(T^{(2)}(X_1,X_1,P)T^{(2)}(X_2,X_2,P)\ \Pi_{j=1}^2 \exp\big(i\xi\ T^{(1)}(X_j,P)\big)\Big)\eta_{n-2}(t)$$

$$+ 2A_n^2 E_P\Big(T^{(2)}(X_1,X_1,P)T^{(2)}(X_1,X_2,P)\ \Pi_{j=1}^2 \exp\big(i\xi\ T^{(1)}(X_j,P)\big)\Big)\eta_{n-2}(t)$$

$$+ 4\ A_n^2 E_P\Big(T^{(2)}(X_1,X_2,P)^2\ \Pi_{j=1}^2 \exp\big(i\xi\ T^{(1)}(X_j,P)\big)\Big)\eta_{n-2}(t)$$

$$+ A_n^1 E_P\Big(T^{(2)}(X_1,X_1,P)^2 \exp\big(i\xi\ T^{(1)}(X_1,P)\big)\Big)\eta_{n-1}(t)\Bigg) ,$$

(A.8.16)
$$S_{4,n}(t) :=$$

$$n^{-3/2}\Bigg(A_n^3 E_P\Big(T^{(3)}(X_1,X_2,X_3,P)\ \Pi_{j=1}^3 \exp\big(i\xi\ T^{(1)}(X_j,P)\big)\Big)\eta_{n-3}(t)$$

$$+ 3\ A_n^2 E_P\Big(T^{(3)}(X_1,X_1,X_2,P)\ \Pi_{j=1}^2 \exp\big(i\xi\ T^{(1)}(X_j,P)\big)\Big)\eta_{n-2}(t)$$

$$+ A_n^1 E_P\Big(T^{(3)}(X_1,X_1,X_1,P) \exp\big(i\xi\ T^{(1)}(X_1,P)\big)\Big)\eta_{n-1}(t)\Bigg) .$$

Let us now calculate the expectations in this expression. For short, put $T_1(j) := T^{(1)}(X_j,P)$, $T_2(j,k) := T^{(2)}(X_j,X_k,P)$ and so on... $\delta > 0$ is a constant which may be chosen arbitrarily small, and until the end of the proof, $P_{j,n}(t)$, $P'_{j,n}(t)$, $j \in \mathbb{N}$, are finite combination of positive powers of t, with bounded coefficients which may depend on n . Remember also that the gradients $T^{(j)}(.,P)$ are orthogonal i.e. if a r.v. appears one time in $T^{(j)}$ and not in $T^{(k)}$, $j\neq k$, then the expectation of the product is 0 .

Using the fact that, for every $p \in \mathbb{N}$, for every $x \in \mathbb{R}$, every $\theta \in [0,1]$,

$$\left|\exp(ix) - \sum_{j=0}^{p} (ix)^j/ j!\right| \le 2 \, |x|^{p+\theta} ,$$

the first term in the r.h.s. of (A.8.14) becomes

$$E_P\left(T_2(1,2)\ \Pi_{j=1}^{2}\exp\left(i\xi T_1(j)\right)\right)$$

$$= E_P\left(T_2(1,2)\ \Pi_{j=1}^{2}\left(1+i\xi T_1(j)-\xi^2/2\ T_1(j)^2+ o(\ \xi\ T_1(j))^{2+\delta}\right)\right)$$

$$= -\xi^2 E_P\left(T_2(1,2)T_1(1)T_1(2)\right) -i\ \xi^3 E_P\left(T_1(1)T_1(2)^2 T_2(1,2)\right) + e_{1,n}$$

with $e_{1,n} := o(n^{-3/2}P_{1,n}(t))$, where the coefficient of $P_{1,n}(t)$ are bounded if the following moments

$$E_P|T_2(1,2)|\,|T_1(1)|^2|T_1(2)|^2 ,\ E_P|T_2(1,2)|\,|T_1(1)|^2|T_1(2)|^{2+\delta} ,$$
$$E_P|T_2(1,2)|\,|T_1(1)|\,|T_1(2)|^{2+\delta},\ E_P|T_2(1,2)|\,|T_1(1)|^{2+\delta}|T_1(2)|^{2+\delta}$$

are defined. But by Cauchy-Schwarz inequality and the independence of $T_1(1)$ and $T_1(2)$ we have for instance

$$E_P\left(|T_2(1,2)|\,|T_1(1)|^{2+\delta}|T_1(2)|^{2+\delta}\right)$$

$$\le E_P\left(T_2(1,2)^2\right)^{1/2}\ E_P\left(|T_1(1)|^{4+2\delta}\right)^{1/2} E_P\left(|T_1(2)|^{4+2\delta}\right)^{1/2}$$

which is finite under (A.8.3) and (A.8.4) provided that δ is chosen small enough ($\delta < \varepsilon_1/2$, for example). The other moments may be handled in the same way.

Similarly, we obtain the second term of the r.h.s. of (A.8.14),

$$E_P\left(T_2(1,1)\exp\left(i\xi\, T_1(1)\right)\right) = E_P\left(T_2(1,1)\right) + i\xi\, E_P\left(T_1(1)T_2(1,1)\right)$$
$$+\, o\left(n^{-1/2}P_{2,n}(t)\right)$$

Let us now consider the expressions in (A.8.15)

$$E_P\left(T_2(1,2)\, T_2(3,4)\, \Pi_{j=1}^4 \exp\left(i\xi\, T_1(j)\right)\right) = \xi^4 \left(E_P\left(T_1(1)T_1(2)T_2(1,2)\right)\right)^2$$
$$+\, o\left(n^{-2}\, P_{3,n}(t)\right)$$

The coefficients in $P_{3,n}$ are functions of

$$E_P|T_1(1)|^i|T_1(2)|^j|T_1(3)|^k|T_1(4)|^l|T_2(1,2)T_2(3,4)|$$

with $(i,j,k,l) \in \{\ 0,1,\ 1+\delta\ \}$. One more time, by Cauchy-Schwarz, all the expectations are finite under (A.8.3) and (A.8.4). We proceed the same way for the other terms in (A.8.15) and obtain

$$E_P\left(T_2(1,1)T_2(3,4)\ \Pi_{j=1}^3\exp\left(i\xi\ T_1(j)\right)\right)$$
$$= -\xi^2 E_P\left(T_2(1,1)\right)E_P\left(T_1(2)T_1(3)T_2(2,3)\right) + o\left(n^{-1}P_{4,n}(t)\right)\ ,$$

$$E_P\left(T_2(1,2)T_2(1,3)\ \Pi_{j=1}^3\exp\left(i\xi\ T_1(j)\right)\right) = -\xi^2\ E_P\left(T_2(1,2)T_2(1,3)T_1(2)T_1(3)\right)$$
$$+\, o\left(n^{-1}P_{5,n}(t)\right)\ ,$$

$$E_P\left(T_2(1,1)T_2(2,2)\ \Pi_{j=1}^2\exp\left(i\xi\ T_1(j)\right)\right) = \left(ET_2(1,1)\right)^2 + o\left(P_{6,n}(t)\right)\ ,$$

$$E_P\left(T_2(1,1)T_2(1,2)\ \Pi_{j=1}^2\exp\left(i\xi\ T_1(j)\right)\right) = o\left(P_{7,n}(t)\right)\ ,$$

$$E_P\left(T_2(1,2)^2\ \Pi_{j=1}^2\exp\left(i\xi\ T_1(j)\right)\right) = E_P\left(T_2(1,2)^2\right) + o\left(P_{8,n}(t)\right)\ ,$$

$$E_P\left(T_2(1,1)^2\exp\left(i\xi\ T_1(1)\right)\right) = E_P\left(T_2(1,1)^2\right) + o\left(P_{9,n}(t)\right)\ ,$$

where the coefficients of the $\left(P_{j,n}\right)$, $1 \le j \le 9$ are all bounded, under (A.8.3) and (A.8.4). Consider now the terms in (A.8.16),

$$E_P\left(T_3(1,2,3)\ \Pi_{j=1}^3 \exp\left(i\xi\ T_1(j)\right)\right) = -i\ \xi^3\ E_P\left(T_3(1,2,3)T_1(1)T_1(2)T_1(3)\right)$$

$$+o\left(n^{-3/2}\ P_{10,n}(t)\right)\ ,$$

$$E_P\left(T_3(1,1,2,P)\Pi_{j=1}^2 \exp\left(i\xi T_1(j)\right)\right) = i\xi E_P T_3(1,1,2)T_1(2)\ +o\left(n^{-1/2}\ P_{11,n}(t)\right)$$

$$E_P\left(T_3(1,1,1,P)\exp\left(i\xi\ T_1(1)\right)\right) = E_P\left(T_3(1,1,1,P)\right) + o\left(P_{12,n}(t)\right)\ ,$$

where the coefficients of the $\left(P_{j,n}\right)$, $10 \le j \le 12$, are all bounded.

iii) Derivation of the principal term in the Edgeworth expansion

Plugging the preceding expectations in (A.8.14)-(A.8.16) and (A.8.13) yields

$$\Psi_n(t) = \eta_n(t) + i\xi/2\ n^{-1}\ A_n^2\ \Big(-\xi^2 E_P\left(T_2(1,2)T_1(1)T_1(2)\right)$$

$$-i\ \xi^3 E_P\left(T_1(1)T_1(2)^2 T_2(1,2)\right)\Big)\eta_{n-2}(t)$$

$$+\ i\xi/2\ n^{-1}\ A_n^1\ \Big(\ E_P\left(T_2(1,1)\right) + i\xi\ E_P\left(T_1(1)T_2(1,1)\right)\Big)\eta_{n-1}(t)$$

$$-\ \xi^2/8\ n^{-2} A_n^4\ \Big(\xi^2 E_P\left(T_1(1)T_1(2)T_2(1,2)\right)\Big)^2\ \eta_{n-4}(t)$$

$$-\ \xi^2/4\ n^{-2} A_n^3\ \Big(-\ \xi^2\ E_P\left(T_2(1,1)\right)E\left(T_1(2)T_1(3)T_2(2,3)\right)\Big)\eta_{n-3}(t)$$

$$-\ \xi^2/2\ n^{-2} A_n^3\ \Big(-\xi^2\ E_P\left(T_2(1,2)T_2(1,3)T_1(2)T_1(3)\right)\Big)\eta_{n-3}(t)$$

$$-\ \xi^2/8\ n^{-2} A_n^2\ \left(E_P T_2(1,1)\right)^2 \eta_{n-2}(t)\ -\ \xi^2/4\ n^{-2} A_n^2\ E_P\left(T_2(1,2)^2\right)\ \eta_{n-2}(t)$$

$$-\ \xi^2/8\ n^{-2} A_n^1\ E_P\left(T_2(1,1)^2\right)\ \eta_{n-1}(t)$$

$$+\ i\xi/6\ n^{-2} A_n^3\ \Big(-i\ \xi^3\ E_P\left(T_3(1,2,3)T_1(1)T_1(2)T_1(3)\right)\Big)\ \eta_{n-3}(t)$$

$$+\ i\xi/2\ n^{-2}\ A_n^2\ \left(i\xi\ E_P T_3(1,1,2)T_1(1)\right)\ \eta_{n-2}(t)$$

$$+\ i\xi/6\ n^{-2} A_n^1\ E_P\left(T_3(1,1,1,P)\right)\ \eta_{n-1}(t)\ +\ r_{3,n}(t)$$

with

$$r_{3,n}(t) := r_{2,n}(t) + r_{4,n}(t)\ ,$$

and

$$r_{4,n}(t) := n^{-3/2} A_n^2 n^{-3/2} P'_{1,n}(t)\, \eta_{n-2}(t)$$

$$+ n^{-3/2} A_n^1 n^{-1/2} P'_{2,n}(t)\, \eta_{n-1}(t)$$

$$+ n^{-3} A_n^4 n^{-2} P'_{3,n}(t)\, \eta_{n-4}(t) + n^{-3} A_n^3 n^{-1} P'_{4,n}(t)\, \eta_{n-3}(t)$$

$$+ n^{-3} A_n^3 n^{-1} P'_{5,n}(t)\, \eta_{n-3}(t) + n^{-3} A_n^2 P'_{6,n}(t)\, \eta_{n-2}(t)$$

$$+ n^{-3} A_n^2 P'_{7,n}(t)\, \eta_{n-2}(t) + n^{-3} A_n^2 P'_{8,n}(t)\, \eta_{n-2}(t)$$

$$+ n^{-3} A_n^1 P'_{9,n}(t)\, \eta_{n-1}(t) + n^{-5/2} A_n^3 n^{-3/2} P'_{10,n}(t)\, \eta_{n-3}(t)$$

$$+ n^{-5/2} A_n^2 n^{-1/2} P'_{11,n}(t)\, \eta_{n-2}(t) + n^{-5/2} A_n^1 P'_{12,n}(t)\, \eta_{n-1}(t)\,.$$

From Gnedenko and Kolmogorov (1968), p.204, we have, for $|t| \le C\, n^{1/2}$, $C > 0$, $1 \le p \le 4$

$$\eta_{n-p}(t) = \eta_n(t) + \exp\left(-t^2/2\right) C(p,P)\, n^{-1} t^2 + |t| \exp(-t^2/4)\, O\left(n^{-1}\right)$$

where $C(p,P)$ is a constant depending on p and P.

Moreover, for $|t| \le C\, n^{1/2}$, $\eta_n(t)$ admits the expansion (see Feller (1971)),

$$\eta_n(t) = \exp(-t^2/2) \Big(1 - it^3 (6\, n^{1/2})^{-1} E_P T_1(1)^3$$

$$+ (24\, n)^{-1}\left(E_P T_1(1)^4 - 3\right)t^4 - (72\, n)^{-1}\left(E_P T_1(1)^3\right)^2 t^6\Big)$$

$$+ Q(t) \exp(-t^2/4)\, O\left(n^{-1}\right)$$

where Q is a polynomial. Therefore we have $e_n(t) = P(t) \exp(-t^2/4)\, O\left(n^{-1}\right)$ where $P(t)$ is a finite linear combination of powers of t, and $O(n^{-1})$ is uniform for $t \in [\,0,\, C\, n^{1/2}]$. It follows that

$$\Psi_n(t) = \exp(-t^2/2) \Big(1 - it^3 (6\, n^{1/2})^{-1} E_P T_1(1)^3$$

$$+ (24\, n)^{-1}\left(E_P T_1(1)^4 - 3\right)t^4 - (72\, n)^{-1}\left(E_P T_1(1)^3\right)^2 t^6$$

$$- i n^{-1/2}\, t^3/2\; E_P\left(T_2(1,2) T_1(1) T_1(2)\right) \left(1 - it^3 (6 n^{1/2})^{-1} E_P T_1(1)^3\right)\Big)$$

$$+ n^{-1} t^4/2\ E_P\left(T_1(1)T_1(2)^2T_2(1,2)\right)\ \exp(-t^2/2)$$

$$+ it/2\ n^{-1/2}\ E_P\left(T_2(1,1)\right)\ \left(1 - it^3(6\ n^{1/2})^{-1}\ E_P T_1(1)^3\right)\exp(-t^2/2)$$

$$- n^{-1} t^2/2\ E_P\left(T_1(1)T_2(1,1)\right)\Big)\exp(-t^2/2)$$

$$- n^{-1} t^6/8\ \left(E_P\left(T_1(1)T_1(2)T_2(1,2)\right)\right)^2 \exp(-t^2/2)$$

$$+ n^{-1} t^4/4\ \left(E_P\left(T_2(1,1)\right)E_P\left(T_1(2)T_1(3)T_2(2,3)\right)\right)\exp(-t^2/2)$$

$$+ n^{-1} t^4/2\ \left(E_P\left(T_2(1,2)T_2(1,3)T_1(2)T_1(3)\right)\right)\exp(-t^2/2)$$

$$- n^{-1} t^2/8\ \left(E_P T_2(1,1)\right)^2\exp(-t^2/2)$$

$$- n^{-1} t^2/4\ E_P\left(T_2(1,2)^2\right)\ \exp(-t^2/2)$$

$$+ n^{-1} t^4/6\ \ E_P\left(T_3(1,2,3)T_1(1)T_1(2)T_1(3)\right)\exp(-t^2/2)$$

$$- n^{-1} t^2/2\ E_P T_3(1,1,2)T_1(2)\Big)\ \exp(-t^2/2)$$

$$+ r_{2,n}(t) + Q(t)\exp(-t^2/4)\ o\left(n^{-1}\right)\ .$$

iv) Control of the remainder

To end the proof, it suffices to control the remainder $r_{2,n}$, given by (A.8.13). By Hölder inequality, we have

$$E_P\left[\Pi_{1,n}(t)\left(\int T^{(2)}(x_1,x_2,P)\ \Pi_{j=1}^2\beta_n(dx_j)\right)\ \left(\int T^{(3)}(x_1,x_2,x_3,P)\ \Pi_{j=1}^3\beta_n(dx_j)\right)\right]$$

$$\le E_P\left(\left|\int T^{(2)}(x_1,x_2,P)\ \Pi_{j=1}^2\beta_n(dx_j)\right|^3\right)^{1/3}$$

$$E_P\left(\left|\int T^{(3)}(x_1,x_2,x_3,P)\ \Pi_{j=1}^3\beta_n(dx_j)\right|^{3/2}\right)^{2/3}$$

and

$$E_P\left(\Pi_{1,n}(t)\left(\int T^{(2)}(x_1,x_2,P)\ \Pi_{j=1}^2\beta_n(dx_j)\right)^2\left(\int T^{(3)}(x_1,x_2,x_3,P)\ \Pi_{j=1}^3\beta_n(dx_j)\right)\right)$$

$$\le E_P\left(\left|\int T^{(2)}(x_1,x_2,P)\ \Pi_{j=1}^2\beta_n(dx_j)\right|^3\right)^{2/3}$$

$$E_P\left(\left|\int T^{(3)}(x_1,x_2,x_3,P)\ \Pi_{j=1}^3\beta_n(dx_j)\right|^3\right)^{1/3}$$

But from Callaert and Janssen (1978) (see also Serfling (1980) chap.5 and 6), we have, under (A.8.4), (A.8.5)

$$E_P\left(\left|\int T^{(2)}(x_1,x_2,P)\ \Pi_{j=1}^2\beta_n(dx_j)\right|^3\right)^{1/3} = O(1)$$

$$E_P\left(\left|\int T^{(3)}(x_1,x_2,x_3,P)\ \Pi_{j=1}^3\beta_n(dx_j)\right|^\zeta\right)^{1/\zeta} = O(1),\ \zeta = 3/2 \text{ or } \zeta = 3.$$

Thus, we deduce

$$r_{2,n}(t) = r_{1,n}(t) + t^2\ O(n^{-3/2}) + t^3\ O(n^{-2}).$$

The remainder term $r_{1,n}(t)$ may be handled similarly because we have

$$|E_P(\Pi_{1,n}(t)\ \ \Pi_{2,n}(n^{-1/2}t)\ \ R_{3,n}(t))| \le E_P|R_{3,n}| = t^{1+a}\ O\left(n^{-(1+a)}\right)$$

$$|E_P(\Pi_{1,n}(t)\ R_{2,n}(t)\ \Pi_{3,n}(n^{-1}t)\ )| \le\ \ E_P|R_{2,n}| = t^{(2+b)}O\left(n^{-(2+b)/2)}\right)$$

$$|E_P(\Pi_{1,n}(t)\ \ R_{2,n}(t)\ R_{3,n}(t))| = t^{3+a+b}\ O\left(n^{-(2+a+b/2)}\right).$$

This gives the bound

$$r_{2,n}(t) = t^{1+a}\ \ O\left(n^{-(1+a)}\right) + t^{(2+b)}\ O\left(n^{-(2+b)/2)}\right)$$

$$+\ t^{3+a+b}O\left(n^{-(2+a+b/2)}\right) + t^2\ O(n^{-3/2}) + t^3\ O(n^{-2}).$$

where the O are uniform over $t\in[0,Cn^{1/2}]$. □

Corollary A.8.3 : **Assume that $S^2(P)$ is second order Fréchet-differentiable, with gradients $S^{(1)}(x,P)$ and $S^{(2)}(x,P)$. Let**

$$\tilde{K}_n(x,P) = P\,(n^{1/2}\,(T(P_n) - T(P))/\,S(P_n) \le x)$$

be the c.d.f. of the studentized statistic, with characteristic function $\psi_n'(t)$. Assume that the conditions of proposition A.8.1 hold with $\psi_n(t)$ replaced by $\psi_n'(t)$. If in addition, we have, for some $\varepsilon > 0$

(A.8.17) $\quad T^{(1)}(.,P) \in L^{(1)}_{9+\varepsilon}(P)$

(A.8.18) $\quad T^{(2)}(.,P) \in L^{(1)}_{9+\varepsilon}(P)$

(A.8.19) $\quad T^{(3)}(.,P) \in L^{(1)}_{3+\varepsilon}(P)$

then $\tilde{K}_n(x,P)$ admits the Edgeworth expansion up to $o(n^{-1})$:

$$\tilde{\mathcal{E}}_{2,K}(x,P) = \Phi\big(x\big) - n^{-1/2}\left(\frac{K'_{1,1}(P)}{2} + \frac{K'_{1,3}(P)}{6}\big(x^2-1\big)\right)\phi(x)$$

$$- n^{-1}\left(\frac{K'_{2,2}(P)}{2}\,x + \frac{K'_{2,4}(P)}{24}\big(x^3-3x\big) + \frac{K'_{2,6}(P)}{24}\big(x^5 - 10x^3 + 15x\big)\right)\phi(x)$$

where the $K'_{i,j}(P)$'s are given by the $K_{i,j}(P)$'s in (A.8.6)-(A.8.10) when $T^{(k)}(.,P)$ is replaced by the gradients $f^{(k)}(.,P)$, $k = 1,..,3$ defined in section II.5.1 (see also Appendix 7).

Proof of proposition A.8.3 : The proof follows from proposition A.8.1 and the calculations of Appendix 7. Just notice that, if we put $f(Q) = T(Q) = T(P))/S(Q)$ then, $\tilde{K}_n(x,P) = P(\,n^{1/2}(f(P_n) - f(P)) \le x)$. The gradients of f are given by the $f^{(k)}$ (see Appendix 7). Thus we just have to check that (A.8.3)-(A.8.5) hold for $f^{(k)}$, which is easy under (A.8.17)-(A.8.20). □

REFERENCES

Abramovitch, L., Singh, K. (1985) : Edgeworth corrected pivotal statistics and the bootstrap, *Ann. Statist.*, 13, 116-132.

Aerts, M., Janssen, P. (1994) : Weighted bootstrapping for U-quantiles, preprint.

Akahira, M., Takeuchi, K.(1991) : Bootstrap method and empirical process, *Ann. Inst. Statist. Math.*, 43, 297-310.

Aldous, D. (1983): *Exchangeability and Related Topics*, in Ecole d'été de Probabilités de St Flours XIII, P.L, Henequin ed., *Lectures Notes in Math.* 1117, Springer Berlin.

Arenal, E., Cuesta-Albertos, J.A., Matran, C. (1994) : On the unconditional strong law of large numbers for the bootstrap mean, preprint.

Athreya, K.B. (1983) : Strong law for the bootstrap, *Statist. Probab. Lett.*, 1, 147-150.

Athreya, K. B. (1987) : Bootstrap mean in the infinite variance case, *Ann. Statist.*, 15, 724-731.

Athreya, K.B., Ghosh, M., Low, L.Y., Sen, P.K. (1984) : Laws of large numbers for bootstrapped U-statistics, *J. Statist. Plann. Inf.*, 9, 185-194.

Arcones, M.A., Giné, E. (1989) : The bootstrap of the mean with arbitrary bootstrap sample size, *Ann. Inst. H. Poincare Probab. Statist.*, 25, 457-481.

M.A., Giné, E. (1992) : On the bootstrap of M-estimators and other statistical functionals, in *Exploring the Limits of Bootstrap*, R. LePage and L. Billard ed., Wiley, New York.

Arcones, M.A., Giné, E. (1993) : On the bootstrap of U and V-statistics, *Ann. Statist.*, 20, 655-674.

Babu, G.J., Singh, K. (1983) : Inference on means using the bootstrap, *Ann. Statist.*, 11, 999-1003.

Bahadur, R.R. (1971) : *Some Limit Theorems in Statistics*, Regional Conference Series on Applied Mathematics, vol. 4, SIAM, Philadelphia, Pensylvania.

Barbe, Ph. (1992a) : Strong laws for the generalized bootstrap mean, preprint.

Barbe, Ph. (1992b) : Bootstrapping the renewal spacing processes, submitted.

Barbe, Ph. (1993) : Weighted approximation of the renewal spacing process, *J. Mult. Anal.*, 45, 171-182.

Barndorff-Nielsen, O.E. (1988) : *Parametric Statistical Models and Likelihood*, Springer, New York.

Barndorff-Nielsen, O.E., Cox, D.R. (1989) : *Asymptotic Techniques for Use in Statistics*, Chapman and Hall, London.

Beran, R. (1984) : Jackknife approximations to bootstrap estimates, *Ann. Statist.*, 12, 101-118.

Beran, R. (1987) : Prepivoting to reduce level error of confidence sets, *Biometrika*, 74, 457-468.

Beran, R. (1988) : Prepivoting test statistics: a bootstrap view of asymptotic refinements, *J. Amer. Statist. Assoc.*, 83, 687-697.

Beran, R., Millar, P.W. (1986) : Confidence sets for a multivariate distribution, *Ann. Statist.*, 14, 431-443.

Beran, R., LeCam, L., Millar, W. (1987) : Convergence of stochastic empirical measures, *J. Mult. Anal.*, 23, 159-168.

Bertail, P. (1992) : *La Méthode du Bootstrap, Quelques Applications et Résultats Théoriques*, Ph. D. dissertation, Université Paris IX.

Bertail, P. (1994) : Second order properties of an extrapolated bootstrap without replacement, under weak assumptions, submitted.

Bhattacharya, R.N. (1987) : Some aspects of Edgeworth expansions in statistics and probability, in *New Perspectives in Applied and Theoretical Statistics*, 157-170, Wiley, New York.

Bhattacharya, R.N., Denker, M. (1990) : *Asymptotic Statistics*, Birkhäuser, Basel

Bhattacharya, R.N., Qumsiyeh, M. (1989) : Second order comparisons between the bootstrap and empirical edgeworth expansions, *Ann. Statist.*, 17, 160-169.

Bhattacharya, R.N., Rao, R. (1986) : *Normal Approximation and Asymptotic Expansions*, Krieger, Melbourn.

Bickel, P.J. (1994): A safer but still accurate bootstrap: extrapolating from lower resample sizes, communication at Chapel Hill IMS meeting.

Bickel, P.J. (1992) : Theoretical comparison of different Bootstrap t confidence bounds, in *Exploring the Limits of Bootstrap*, R. LePage and L. Billard ed., Wiley, New York.

Bickel, P.J., Freedman, D.A. (1980) : On Edgeworth expansion and the Bootstrap, unpublished manuscript.

Bickel, P.J., Freedman, D.A. (1981) : Some asymptotic theory for the bootstrap, *Ann. Statist.*, 9, 1196-1217.

Bickel, P.J., Götze, F, van Zwet, W.R. (1986) : The Edgeworth expansion for U-statistics of degree 2 , *Ann. Statist.*, 14, 1463-1484.

Bingham, N.H., Goldie, C.M., Teugels, J. (1987) : *Regular Variation*, Cambridge University Press, Cambridge.

Birman, M. Š., Solomjak, M.Z. (1967) : Piecewise polynomial approximations and functions for the class W_p^α , *Math. Sb.* , 73, 295-317.

Boos, D., Monahan, J.F. (1986) : Bootstrap methods using prior information, *Biometrika*, 73, 77-83.

Bretagnolle, J. (1983) : Lois limites du Bootstrap de certaines fonctionnelles, *Ann. Inst. H. Poincaré Statist. Probab.*, XIX, 281-296.

Broniatowski, M. (1992) : Tauberian theorems, Chernoff inequality and the tail behaviour of finite convolutions of distribution functions, preprint.

Bürke, M.D., Csörgő, M. (1976) : Weak approximation of empirical process when parameters are estimated, in *Empirical Distributions and Processes*, P.

Gaenssler and P. Révész eds., *Lecture Notes in Mathematics*, 566, 1-16, Springer Verlag, Berlin.

Burke, M.D., Csörgő, M., Csörgő, S., Révész, P. (1979) : Approximation of the empirical process when the parameters are estimated, *Ann. Probab.*, 7, 790-810.

Burke, M.D., Gombay, E. (1988) : On goodness of fit and the bootstrap, *Statist. Probab. Letters*, 6, 287-293.

Callaert, H., Janssen, P. (1978) : The Berry-Esseen theorem for U-statistics, *Ann. Statist.*, 6, 417-421.

Callaert, H., Janssen, P., Veraverbeke, N. (1980) : An Edgeworth expansion for U-statistics, *Ann. Statist.*, 8, 299-312.

Chaganty, N.R., Karandikar, R.L. (1992) : Large deviations for the Bootstrap empirical measure, Tech. rep. TR92-15, Dept of Mathematics and Statistics, Old Dominion University.

Chernoff, H. (1952) : A measure of asymptotic efficiency for tests of a hypothesis based on the sum of observations, *Ann. Math. Statist.*, 23, 493-507.

Chibishov, D. M. (1972) : An asymptotic expansion for the distribution of a statistic admitting an asymptotic expansion, *Theory Probab. Appl.*, 17, 620-630.

Chow, Y., Teicher, H. (1988) : *Probability Theory, Independence, Interchangeability, Martingales*, Springer, New York.

Csörgo, S. (1991) : On the law of large numbers for the bootstrap mean, *Statist. Probab. Lett.*, 14, 1-7.

Csörgo, M., Csörgo, S., Horváth, L., Mason, D.M. (1986a) : Weighted empirical and quantile processes, *Ann. Probab.*, 14, 31-85.

Csörgo, M., Csörgo, S., Horváth, L., Mason, D.M. (1986b) : Normal and stable convergence of integral functions of the empirical distribution function, *Ann. Probab.*, 14, 31-85.

Csörgo, S., Mason, D. M. (1989) : Bootstrapping empirical functions, *Ann. Statist.*, 17, 1447-1471.

Csörgo, M., Révész, P. (1981) : *Strong Approximation in Probability and Statistics*, Academic Press, New York.

De Haan, L. (1970) : *On Regular Variation and its Application to the Weak Convergence of Sample Extremes*, Mathematical Centre Tract, vol. 32, Mathematisch Centrum, Amsterdam.

Deheuvels, P., Mason, D.M., Shorack, G.R. (1993) : Some results on the influence of extremes on the bootstrap, *Ann. Inst. Henri Poincaré, Probab. Statist.*, 29, 83-103.

Dehling, H., Denker, M., Woyczynski, W.A. (1990) : Resampling U-statistics using p-stable laws, *J. Mult. Anal.*, 34, 1-18.

Dehling, H., Mikosch, T. (1992) : Random quadratic forms and the bootstrap of U-statistics, preprint.

Dembo, A., Zeitouni, O. (1993) : *Large Deviation Techniques and Applications*, Jones and Bartlett, Boston.

Devroye, L. (1986) : *Non-uniform Random Variate Generation* , Springer Verlag, New-York

Diaconis, P., Efron B. (1983) : Methodes de Calculs Statistiques Intensifs sur Ordinateurs, *Pour la Science* (Translation of *The American Scientist*)

Diaconis, P., Freedman, D. (1987) : A dozen of de Finetti-style results in search of a theory, *Ann. Inst. Henri Poincaré, Probab. Statist.*, 23, 397-423.

DiCiccio, T.J., Hall, P., Romano, J.P. (1989) : On smoothing the bootstrap, *Ann. Statist.*, 692-704.

DiCiccio, T.J., Romano, J. (1988) : A review of bootstrap confidence intervals (with discussions), *J. Roy. Statist. Soc.*, ser. B, 50, 338-370.

DiCiccio, T.J., Tibshirani, R. (1987) : Bootstrap confidence intervals and bootstrap approximations, *J. Amer. Statist. Assoc.*, 82, 163-170.

Dohmann, B. (1989) : *Confidence Intervals for Quantiles Based on Small Sample, Bootstrap vs Standard Method*, PhD thesis, univ. of Siegen.

Dudley, R.M. (1974) : Metric entropy of some classes of sets with differentiable boundaries, *J. Approx. Theor.*, 10, 227-236.

Dudley, R.M. (1984) : *A Course On Empirical Process*, Ecole d'été de Saint-Flour, *Lecture Notes in Math.*, 1097, Springer-Verlag, New-York.

Dudley, R.M. (1987) Universal donsker classes and metric entropy , *Ann. Probab.*, 15, 1306-1326.

Dudley, R.M. (1990) : Non linear functionals of empirical measures and the bootstrap, in *Probability in Banach Spaces*, 7, E. Eberlein, J Kielbs, M.B. Marcus eds., Birkhauser, Boston.

Dudley, R.M. (1992) : Fréchet differentiability, p-variation and uniform Donsker classes, *Ann. Probab.*, 20, 1968-1982.

Dudley, R.M. (1994) : The order of the remainder term in derivatives of composition and inverse operators for p-variation norms, *Ann. Statist.*, 22, 1-20.

Durbin, J. (1973a) : *Distribution Theory for Test Based on the Sample Distribution Function*, Regional Conference Series in Applied Mathematics, vol.9, SIAM, Philadelphia.

Durbin, J. (1973b) : Weak convergence of the sample distribution when parameters are estimated, *Ann. Statist.*, 1, 279-290.

Dvoretzky, A., Kiefer, J., Wolfowitz, J. (1956) : Asymptotic minimax character of the sample distribution function of the classical multinomial estimator, *Ann. Math. Statist.*, 27, 642-669.

Efron, B. (1979) : Bootstrap methods: another look at the jackknife, *Ann. Statist.*, 7 n°1, 1-26.

Efron, B. (1981) : Nonparametric estimates of standard errors: the jackknife, the bootstrap and other methods, *Biometrika*, 68, 589-599.

Efron, B. (1982) : *The Jackknife, the Bootstrap, and Other Resampling Plans*, CBMS-NF n°38, S.I.A.M., Philadelphia.

Efron, B. (1985) : Bootstrap confidence intervals for a class of parametric problems, *Biometrika*, 42-58.

Efron, B. (1987) : Better bootstrap confidence intervals (with discussions), *J. Amer. Statist. Assoc.*, 82, 171-200.

Efron, B., Gong, G. (1983) : A leisurely look at the bootstrap, the jackknife and cross validation, *Amer. Statist.*, 37, 36-48.

Efron, B., Tibshirani, R. (1986) : Bootstrap methods for standard errors, confidence intervals and other measures of statistical accuracy, *Statist. Sci.*, 1, 54-77.

Efron, B., Tibshirani, R. (1993) : *An Introduction to the Bootstrap*, Chapman and Hall.

Einmahl, U., Mason, D. (1992) : Approximations to permutation and exchangeable process, *J. Theor. Probab.*, 5, 101-126

Falk, M. (1988) : Weak convergence of the bootstrap process for large quantile, *Statist. Decisions*, 6, 385-396.

Falk, M., Reiss R.D. (1989) Weak convergence of smoothed and non-smoothed bootstrap quantiles estimates, *Ann. Probab.*, 362-371.

Falk, M., Kaufmann, E. (1991) Coverage probabilities of bootstrap confidence intervals for quantiles, *Ann. Statist.*, 19, 485-495.

Fernholz, L.T. (1983) : *Von Mises Calculus for Statistical Functionals, Lecture Notes in Statist.*, 19, Springer Verlag, New-York.

Feller, W. (1971) : *An Introduction to Probability Theory and Its Applications*, vol.2, 2nd edition, Wiley, New York.

Field, C., Ronchetti, E. (1990) : *Small Sample Asymptotics*, IMS Lecture Notes, Hayward, CA.

Filippova, A.A. (1961) : Mises' theorem and the asymptotic behaviour of functionals of empirical distribution function and its statistical applications, *Theor. Probab. Appl.*, 7, 25-57.

Flett, T.M. (1980) : *Differential Analysis*, Cambridge university press, London.

Gaenssler, P. (1986) : Bootstrapping empirical measures indexed by Vapnik-Chervonenkis classes of sets, *Probab. Theory Math. Statist.*, 467-481.

Gaenssler, P., Stute, W. (1987) : *Seminar on Empirical Process*, Birkhauser Verlag, Basel-Boston.

Gallambos, J. (1978) : *The Asymtotic Theory of Extreme Order Statistics*, Wiley, New York.

Ghosh, J.K. (1994) : *Higher Order Asymptotics*, NSF-CBMS, Reg. Conf. Ser. Prob. Statist., vol. 4, IMS-ASA.

Ghosh, M., Parr, W.C., Singh, K., Babu, G.J. (1984) : A note on bootstrapping the sample median, *Ann. Statist.*, 12, 1130-1135.

Gill, R.D. (1989) : Non and semi-parametric maximum likehood estimators and the Von Mises method , *Scand. J. Statist.*, 16, 97-128.

Giné, E., Zinn, J. (1989) : Necessary conditions for the bootstrap of the mean, *Ann. Statist.*, 17 n°2., 684-691.

Giné, E., Zinn, J. (1990) : Bootstrapping general empirical functions, *Ann. Probab.*, 18, 851-869.

Giné, E., Zinn, J. (1991) : Gaussian characterization of uniform Donsker classes of functions, *Ann. Probab.*, 19, 758-782.

Gnedenko, B.V., Kolmogorov, A.N. (1968) : *Limit Distribution for Sums of Independent Random Variables*, Addison-Wesley.

Götze, F. (1979) : Asymptotic expansion for bivariate Von Mises functional, *Zeit. Warhsch. Verw. Geb.*, 50, 333-355.

Götze, F. (1984) : Expansions for Von Mises functionals, *Zeit. Warhsch. Verw. Geb.*, 65, 599-625.

Groeneboom, P., Oosterhoff, J., Ruymgaart, F.H. (1979) : Large deviation theorems for empirical probability measure, *Ann. Probab.*, 7, 553-586

Haeusler, E. (1992) : coverage probability of generalized bootstrapped confidence intervals for the mean, manuscript.

Haeusler, E., Mason, D.M., Newton, M.A. (1992) : Weighted bootstrapping of means, *CWI Quaterly*, 213-228.

Härdle, W. (1989) : Resampling from Inference Curve, 47th Proc I.S.I.

Hahn, M.G., Mason, D.M., Weiner, D.C. eds. (1991) : *Sums, Trimmed Sums and Extremes*, Birkhauser, Boston.

Hall, P. (1983) : Inverting an Edgeworth expansion, *Ann. Statist.*,11, 569-576.

Hall, P. (1986) : On the bootstrap and confidence intervals, *Ann. Statist.*, 14, 1431-1452.

Hall, P. (1988a) : Rate of convergence in bootstrap approximations, *Ann. Statist.*, 16, 1665-1684.

Hall, P. (1988b) : Theoretical comparison of bootstrap confidence intervals, *Ann. Statist.*, 16, 927-985.

Hall, P. (1990) : Asymptotic properties of the bootstrap for heavy tails distributions, *Ann. Statist.*, 18, 1342-1360.

Hall, P. (1991a) : Bahadur representations for uniform resampling and importance resampling, with applications to asymptotic relative efficiency, *Ann. Statist.*, 19, 1062-1072.

Hall, P. (1992) : *The Bootstrap and Edgeworth Expansion*, Springer, New-York

Hall, P. Mammen, E. (1992) : A general resampling algorithm and their performance in distribution estimation, preprint.

Hall, P., Martin, M. (1987) : Exact convergence rate of bootstrap quantile variance estimator, *Prob. Theor. Rel. Fields*, 80, 261-268.

Hall, P., Martin, M. (1989) : A note on the accuracy of bootstrapped percentile method confidence intervals for a quantile, *Statist. Probab. Let.*, 8, 197-200.

Hall, P., Martin, M. (1991) On the error incurred using the bootstrap variance estimate when constructing confidence intervals, *J. Mult. Anal.*, 38, 70-81.

Hampel, F. (1974) : The influence curve and its role in robust estimation, *J. Amer. Statist. Assoc.*, 69, 383-393.

Härdle, W., Marron, S. (1991) : Bootstrap simultaneous error bars for non-parametric regression, *Ann. Statist.*, 19, 778-796.

Helmers, R. (1991) : On the Edgeworth expansion and the bootstrap approximation for a studentized U-statistics, *Ann. Statist.*, 9, 470-484.

Helmers, R., Janssen, P., Veraverbeke, N. (1992) : Bootstrapping U-quantiles in *Exploring the Limit of the Bootstrap*, R. Le Page and L. Billiard ed., Wiley, New York, 145-155.

Hinkley, D.V. (1988) : Bootstrap methods, *J. Roy. Statist. Soc., Ser. B*, 50, 321-337.

Hoeffding, W. (1963) : Probability inequalities for sums of bounded random variables, *J. Amer. Statist. Assoc.*, 58, 13-30.

Hoffmann-Jørgensen J. (1991) : *Stochastic Processes on Polish Spaces*, Aarhus universitet.

Huber, P.J. (1981) : *Robust Statistics*, Wiley, New-York.

Hušková, M., Janssen, P. (1993a) : Generalized bootstrap for studentized U-statistics : a rank statistics approach, *Statist. Probab. Let.*, 16, 225-233.

Hušková, M., Janssen, P. (1993b) : Consistency of the generalized bootstrap for degenerate U-statistics, *Ann. Statist.*, 21, 1811-1823

James, L.F. (1993) : *The Bootstrap, Bayesian Bootstrap and Random Weighted Methods for Censored Data Models*, Ph. Dissertation, SUNY at Buffalo.

Janssen, P. (1993) : Weighted bootstrapping of U-statistics, *J. Statist. Plann. Infer.*, 34,

Kiefer, J. (1972) : Iterated logarithm analogue for sample quantiles when $p_n \to 0$, *Proc. Sixth Berkeley Symp. Math. Statist. Probab.*, vol 1, pp.227-244, University of California Press, Berkeley, California.

Knight, K. (1989) : On the bootstrap of the sample mean in the infinite variance case, *Ann. Statist.*, 17, 1168-1175.

Kolmogorov, A.N., Tihomirov, V.M. (1959) : ε-entropy and ε-capacity of sets in functional spaces, *Amer. Math. Soc. Transl.* (ser. 2), 17(1961) (from Uspekhi Mat. Nauk. 14, 3-86)

Konishi, S. (1991) : Normalizing transformations and bootstrap confidence intervals, *Ann. Statist.*, 19, 2209-2225.

Künsch, H.R. (1989); The jackknife and the bootstrap for general stationnary observations, *Ann. Statist.*, 17, 1217-1241.

Liu, R., Singh, K. (1987): On partial correction by bootstrap, *Ann. Statist.*, 15, 1713-1718.

LePage, R., Billiard, L. ed. (1992) : *Exploring the Limits of the Bootstrap*, Wiley, New York.

Lo, A. Y. (1987) : A large sample study of the Bayesian bootstrap, *Ann. Statist.*, 15, 360-375.

Lo, A. Y. (1991) : Bayesian bootstrap clones and a biometry function, *Sankhya, Ser. A*, 53, 320-333

Lohse, K. (1984) : *Zur Konsistenz des Bootstrap-Verfahrens*, Ph. D. thesis, University of Hamburg.

Lohse, K. (1987) : Consistency of the bootstrap, *Statist. Decisions*, 5, 353-366.

Lorentz, G.G. (1986) : *Approximation of Functions*, 2nd Ed., Chelsea, New York

Mammen, E. (1992) : *When Does Bootstrap Work : Asymptotic Results and Simulations*, Lecture Notes in Statistics, 77, Springer, Berlin.

Mammen, E. (1993) : Bootstrap and wild bootstrap for high dimensional linear models, *Ann. Statist.*, 21, 255-285.

Manski, C.F. (1988) : *Analog Estimation Methods in Econometrics*, Chapman and Hall, New-York.

Mason, D.M. (1991) : A note on weighted approximations to the uniform empirical and quantile processes, in *Sums, Trimmed Sums and Extremes*, M.G. Haan, D.M. Mason, D.G. Weiner Eds., Birkhäuser, Boston.

Mason, D.M., Newton, M. A. (1992) : A rank statistic approach to the consistency of a general bootstrap, *Ann. Statist.*, 20, 1611-1624.

Mason, D.M., Van Zwet, W.R. (1987) : A refinement of the KMT inequality for the uniform empirical process, *Ann. Probab.*, 15, 871-884.

Parr, W.C. (1985) : The bootstrap: some large sample theory and connections with robustness, *Statist. Probab. Lett.*, 3, 97-100.

Pfanzagl, J. (1985) : *Asymptotic Expansion for General Statistical Models*, *Lecture Notes in Statist.*, 31, Springer-Verlag, Berlin.

Pfanzagl, J. (1990) : *Estimation in Semiparametric Model*, *Lecture Notes in Statist.*, 63, Springer-Verlag, Berlin.

Politis, D.N., Romano, J.P. (1992) : A general theory for the large sample confidence regions based on subsamples under minimal assumption, Tech. Rep. 399, Dept. Statist., Stanford University.

Pollard, D. (1984) : *Convergence of Stochastic Processes*, Springer Verlag New-York.

Pons, O., Turkheim, E. (1989) Methodes de Von Mises, Hadamard différentiabilité et bootstrap dans un modèle non paramétrique sur un espace métrique, *C. R. Acad. Sci. Paris*, 308, 369-372.

Praestgaard, J. (1990) : Bootstrap with general weights and multiplier central limit theorems, Technical Report #195 Department of Statistics, University of Washington.

Praestgaard, J. (1992) : General weights bootstrap of the empirical process, Ph. D. dissertation, Univ. of Washington.

Praestgaard, J., Wellner, J.A. (1993) : Exchangeably weighted bootstrap of the general empirical process, *Ann. Probab.*, 21, 2053-2086.

Quenouille, M.H. (1949) : Approximate tests of correlation in time-series, *J. Roy. Statist. Soc., Ser. B*, 11, 68-84.

Rachev, S.T. (1991) : *Probability Metrics and the Stability of Stochastic Models*, Wiley, New York.

Reeds, J.A. (1976) : *On the Definition of Von Mises Functional*, Ph. D. Thesis, Dept Statist., Harvard University.

Reid, N. (1988) : Saddlepoint method and statistical inference, *Statist. Sci.*, 3, 213-227.

Rockafellar, R. (1970) : *Convex Analysis*, Princeton University Press, Princeton, New Jersey.

Rubin, D. (1981) : The Bayesian bootstrap, *Ann. Statist.*, 9, 130-134.

Rudin, W. (1974) : *Real and Complex Analysis*, 2nd Edition, Mc Graw-Hill, New-York.

Sen , P.K. (1988) : Functional approaches in resampling plans: a review of some recent developments, *Sankhya (A)*, 50, 394-435.

Serfling, J. (1980) : *Approximation Theorem of Mathematical Statistics*, Wiley, New-York.

Shao, J. (1989) : The efficiency and consistency of approximations to the jackknife variance estimators, *J. Amer. Statist. Assoc.*, 84, 114-119.

Shao, J. (1991) : Second order differentiability and Jackknife, *Statist.Sinica*, 1, 185-202.

Shao, J., Wu, C.F.J. (1989) : A general theory for jackknife variance estimation, *Ann. Statist.*, 17, 1176-1197.

Sheehy, A., Wellner, J.A. (1992) : Uniform Donsker classes of functions, *Ann. Probab.*, 20, 1983-2030.

Shorack, G.R. (1982) : Bootstrapping robust regression, *Comm. Statist., Theory Methods*, 11 n°9, 961-972.

Shorack, G.R., Wellner, J.A. (1986) : *Empirical Processes with Applications to Statistics*, Wiley, New York.

Silverman B., Young G.A. (1987): The bootstrap : to smooth or not to smooth, *Biometrika*, 74, 469-479.

Singh, K. (1981) : On the asymptotic accuracy of Efron's bootstrap, *Ann. Statist.*, 9, 1187-1195.

Skorohod, A.V. (1976) : On a representation of random variables, *Theor. Probab. Appl.*, 21, 628-631.

Swanepoel, J.W.H. (1986) : A note proving that the (modified) bootstrap works, *Comm. Statist., Theor. Methods*, 15, 1399-1415.

Tu, D.S. (1992) : Approximating the distribution of a general standardized functional statistic with that of jackknife pseudo values, in *Exploring the Limits of Bootstrap*, R. LePage and L. Billard ed., Wiley, New York.

Tu, D.S., Zheng, Z. (1988) : Random weighting method in regression model, *Scienta Sinica A*, XXXI, 1442-1459.

Von Mises, R. (1947) : On the asymptotic distribution of differential functions, *Ann. Math. Statist.*, 18, 309-348.

Withers, C.S. (1983) : Expansions for the distribution and the quantiles of a regular functional of the empirical distribution with applications to confidence intervals, *Ann. Statist.*, 11, 567-587.

Wellner, J.A. (1991) : Empirical processes in action : a review, 48th Proc. I.S.I. Session, Cairo.

Weng, C.S. (1989) : On a second-order asymptotic property of the Bayesian bootstrap mean, *Ann. Statist.*, 17, 705-710.

Wu, C.F.J. (1986) : Jackknife, bootstrap and other resampling methods in regression analysis (with discussions), *Ann. Statist.*, 29, 635-654.

Yamamuro, H. (1974) : *Differential Calculus in Topological Linear Spaces, Lecture Note in Math.*, 374, Springer-Verlag, New-York.

Young, G.A. (1988): A note on bootstrapping the correlation coefficient, *Biometrika*, 75, 370-373.

Zhidong, B., Lincheng, Z. (1986) : Edgeworth expansion of distribution function of independent random variables, *Scientia Sinica (ser.A)*, 29, 1-22.

NOTATION INDEX

$a = E_P T^{(1)}(X,P)^3$ accelerator, 27

$a_n(I,J)$ estimator of a , 28

$\alpha_n(x)$ uniform empirical process, 145

α_n a normalizing sequence, 86, 89

$\hat{\alpha}_n = Q_n(1-2m_n^{-1}) - Q_n(1-m_n^{-1})$, 90

B separable Banach space , 9

$\beta_X = \mathrm{Skew}(X_1)$, skewness of X_1 50

$\beta_{W,n} = E(nW_{1,n} - 1)^3/\sigma_{W,n}^3$ skewness of $nW_{1,n}$, 25

$\hat{\beta}_{X,n} = n^{-1} \Sigma_{i=1}^n (X_i-\overline{X}_n)^3 / \hat{\sigma}_n^3$ empirical skewness of the X_i's, 49

$\beta_{Y,n} = \mathrm{Skew}(Y_{1,n})$, 49

$\beta_{W,n}^{(j)} = \mathrm{Skew}(n\, W_{i,n}^{(j)})$, 33

$\beta_{Y,n}^{(j)} = \mathrm{Skew}(Y_{i,n}^{(j)})$, 33

$B(t)$, $B_n(t)$ Brownian bridges , 79

β_n normalizing sequence , 88

$B_{2,n}(x)$ Edgeworth expansion of $H_n^T(x)$, 51

$B_{2,n}^Y(x)$ Edgeworth expansion of $H_{Y,n}^T(x)$, 51

c_n normalizing sequence, 87

$c_{Y,\alpha,n}$ α-quantile of $G_{Y,n}$, 52

$c_{W,n}(\alpha)$ approximation of $\gamma_{W,n}(\alpha)$, 61

$d(P,Q)$ distance between P and Q , 13

$d_{\mathcal{H}}(P,Q) = \sup_{h\in\mathcal{H}} \left|\int h\, d(P-Q)\right|$ metric induced by $\mathcal{H}$, 14

D dimension of the class $\mathcal{H}$, 17

$DT_P^{(j)}(.)$ differential of order of T at P , 11

$D_{W,n} = n^{1/2}(T(P_{W,n}) - T(P_n))$, 55

$\mathbb{D}_n(x) = n^{-1} \Sigma_{i=1}^n \mathbb{I}\{ U_i \le g_n(x) \}$ perturbated , uniform empirical process , 78

$\mathbb{D}_n^{\leftarrow}(x) = \inf \{ t : \mathbb{D}_n(t) \ge x \}$ generalized inverse of $\mathbb{D}_n(x)$, 79

$\mathbb{D}_{W,n}(x) = \Sigma_{i=1}^n W_{i,n}\, \mathbb{I}\{ U_i \le g_{W,n}(x)\}$ generalized bootstrap version of $\mathbb{D}_n$, 80

δ_{X_i} Dirac measure at X_i , 9

$\Delta_{Y,n} = 2\sigma_{Y,n}(\sigma_{Y,n} - \beta_{Y,n}) + \sigma_{Y,n}^2 - 1$, 63

$\Delta K_{i,j}^{Y,n}(P)$, 64

$\Delta K_{i,j}^{\prime Y,n}(P)$, 68

$\varepsilon^{(m)}(.,P)$ some bound for the remainder term $R^{(m)}(.,.)$, 13

E_W expectation under W_n conditioned on the sample , 10

$\mathcal{E}_{2,H}(x,P)$ Edgeworth expansion of $H_n(x,P)$ up to $o(n^{-1})$, 54

$\mathcal{E}^W_{2,H}(x,P)$ Edgeworth expansion for $H_{W,n}(x)$ up to $o(n^{-1})$, 57

$\mathcal{E}_{2,K}(x,P)$ Edgeworth expansion of $K_n(x,P)$ up to $o(n^{-1})$, 63

$\mathcal{E}_{2,G}(.)$ Edgeworth expansion for $G_{Y,n}$, 50

$f(Q) = (T(Q)-T(P))/\tilde{S}(Q)$, 62

$f^{(j)}(.,P)$ j-th gradient of $f(Q)$, 32

$F_n(x) = n^{-1}\Sigma_{i=1}^n \mathbb{I}(X_i \le x)$ empirical c.d.f. , 83

$F_n^*(x)$ Efron's bootstrap c.d.f. , 6

$F_{W,n}(x) = \Sigma_{i=1}^n W_{i,n}\mathbb{I}\{X_i \le x\}$ generalized boostrap c.d.f. , 84

$\hat{F}_n$ estimator of F , 3

$\tilde{F}_n(x)$ empirical c.d.f. with estimated parameters , 81

ϕ density of $N(0,1)$, 24

Φ c.d.f. of $N(0,1)$, 24

$(g_n)_{n\ge1}$ sequence of functions in the perturbated uniform empirical process, 78

$\tilde{g}_n$ deterministic process approximating g_n , 79

$g_{W,n}$ generalized bootstrap analogue of g_n , 80

$G_n(x) = P\{\, n^{1/2}(\overline{X}_n - EX_1)/\sigma \le x \,\}$ c.d.f. of the mean, 49

$G_{Y,n}(x) = P\{\, n^{1/2}(\overline{X}_{Y,n} - \overline{X}_n\overline{Y}_n)/\hat{\sigma}_n\sigma_{Y,n} \le x \mid P_n \,\}$ c.d.f. of the generalized bootstrap mean, 49

$\gamma_{W,n}(\alpha)$ be the α-quantile of the generalized bootstrapped distribution, 59

$\gamma(x)$ Chernoff transform of L , 74

$\gamma(B)$ limiting process of $n^{1/2}(\tilde{g}_n - g_n)$, 80

$\Gamma(B,x)$ limiting distribution for an empirical process with estimated parameters, 84

$\mathcal{H}$ class of real valued functions on B 14

H envelope of $\mathcal{H}$, 15

$H_n(x,P) = P\{ n^{1/2}(T(P_n)-T(P))/S(P) \le x \}$, 23

$H_{W,n}(x) = P\{ n^{1/2}(T(P_{W,n}) - T(P_n))/(\sigma_{\tau,n}\sigma_{Y,n}) \le x \mid P_n \}$, 57

$H_n^T(x) = P\{ n^{1/2}|\bar{X}_n - EX_1|/\sigma \le x \}$, 51

$H_{Y,n}^T = P\{ n^{1/2}|\bar{X}_{Y,n} - \bar{X}_n|/(\sigma_{Y,n}\sigma_n) \le x \}$, 51

$H_c(x,y,w,z)$, $H_0(x,y,w,z)$, 89

(h,g,k) parameters of the moments of the i.i.d. weights, 52, 70

I(P) efficiency bound for S(P) , 22

$J(Q,P) = \int \gamma(dQ/dP)dP$ if $Q \ll P$, large deviation functional, 74

$Kur(Y) = E(Y-EY)^4/Var(Y)^2$ kurtosis of Y , 33

$K_X = Kur(X_1)$, 50

$K_{Y,n} = E(Y_{i,n} - 1)^4/\sigma_{Y,n}^4$ kurtosis of $Y_{i,n}$, 49

$\hat{K}_{X,n} = n^{-1} \Sigma_{i=1}^n (X_i-\bar{X}_n)^4 / \hat{\sigma}_n^4$ empirical kurtosis , 50

$K_n(x,P) = P\{ n^{1/2}(T(P_n)-T(P))/S_n \le x \}$, 23

$\tilde{K}_n(x,P)$, 137

$K_{i,j}(P)$ coefficient in the Edgeworth expansion of $H_n(.,P)$ up to $o(n^{-i/2})$, 24, 54

$K'_{i,j}(P)$ coefficient in the Edgeworth expansion of $K_n(.,P)$ up to $o(n^{-i/2})$, 24, 63

$K_{1,n}$ estimator of $K_{1,1}(P)$, 25

$K_{3,n}$ estimator of $K_{1,3}(P)$, 25

$K'_{1,n}$ estimator of $K'_{1,1}(P)$, 29

$K'_{3,n}$ estimator of $K'_{1,3}(P)$, 29

$K^Y_{i,j,n}(P)$ coefficients in $\mathcal{E}^W_{2,H}(x)$, 57

$K^{Y,n}_{i,j}(P)$ coefficient in Edgeworth expansion $\mathcal{E}^W_{2,H}(x)$, 64

$K^{Y,n(1)}_{1,1}(x,P)$, $K^{Y,n(1)}_{1,3}(x,P)$ influence function of $K^{Y,n}_{1,1}(P)$ and $K^{Y,n}_{1,3}(P)$, 65

K(Q,P) Kullback-Leibler information number between Q and P , 73

$L(t) = E\exp(tY_{i,n})$ moment generating function , 74

$L_p^{(k)}(P)$, 183

$\mathcal{L}_\alpha$ stable distribution with index $0 < \alpha < 2$, 90

λ exponent bound $\varepsilon^{(2)}(t,P) \le K_0(P)t^\lambda$ $(t \to 0)$, 21

m location parameter, 78

$m(t) = d(\log L(t))/dt$, 74

m_n undersampling size, 1

$m^{(j)}_{W,n} = E_W n\, W^{(j)}_{1,n}$, 33

$m^{(j)}_{y,n} = E_W Y^{(j)}_{1,n}$, 33

$MZ(Z_i,s)$ constant in the Marcinkiewicz-Zygmund inequality, 123

μ exponent bound $\varepsilon^{(2)}(t,P) \le K_\infty(P)t^{\mu}$ $(t \to \infty)$, 20

μ_n estimator for the location scale parameter , 78

$\mu_{p,n} = n^{1/2}\, E_W(n^{1/2}(T(P_{W,n})-T(P_n)))^p$, 25

$N(t)$ Poisson process, 88

$N_1(\varepsilon,Q,\mathcal{H})$ ε-covering number of $\mathcal{H}$, 17

$\| \;\|$ sup norm, 5

$\|f\|_n = \sup_{t\in[1/n,1-1/n]} |f(x)|$, 79

$\| \;\|^{(k)}_{p,P}$ norm, 183

$\omega_1, .., \omega_n$ i.i.d exponential r.v.'s , 85

$\Omega_k = \Sigma^k_{i=1}\, \omega_i$, 85

Ω subset of $\mathcal{P}$, 72

$\mathcal{P}$ space of probability measures, 9

$P_n = n^{-1}\, \Sigma^n_{i=1}\delta_{X_i}$ empirical p.m., 1, 9

$P^*_{m_n} = m_n^{-1}\Sigma^n_{i=1}\delta_{X^*_i}$, 1, 73

$P_{W,n} = \Sigma^n_{i=1}\, W_{i,n}\delta_{X_i}$ generalized bootstrap empirical p.m., 4, 9

$P_{Y,n} = n^{-1}\Sigma^n_{i=1}\, Y_{i,n}\delta_{X_i}$, 74

$p_n(\alpha) = P\{\, n^{1/2}(\overline{X}_n - EX_1)/\sigma \le c_{Y,\alpha,n} \}$ coverage probability for the mean , 52

$p^H_{c,n}(\alpha) = P\{\, n^{1/2}(T(P_n) - T(P))/S(P) \le \gamma_{W,n}(\alpha) \}$, 64

$p^K_{c,n} = P\{\, n^{1/2}(T(P_n) - T(P))/\, S_n \le \gamma_{n,W}(\alpha) \}$, 66

$\Pi^{Y,n(1)}_2(x,P)$ complicated influence function! 65

$\Pi_n(t)$, $\Pi^i_{n-1}(t)$, $\Pi^{i,j}_{n-2}(t)$ some caracteristic functions, 129, 130

$\psi_{Y,n}$ = characteristic function of the quadratic part of $n^{1/2}(T(P_{w,n}) - T(P_n))$ for i.i.d. random weights, 128

$\Pi_i(P)$, i=1,2 functionals, 138, 142

$Q(.) = F^{\leftarrow}(.)$ quantile function, 89

$Q_n(\ ) = F_n^{\leftarrow}(\ )$ empirical quantile function, 89

$Q_{W,n}(\ ) = F_{W,n}^{\leftarrow}(\ )$ generalized bootstrap quantile function, 89

$Q_{W,n}$ quadratic part of $n^{1/2}(T(P_{W,n}) - T(P_n))$, 99

Q_n quadratic part of $n^{1/2}(T(P_n) - T(P))$, 108

$R^{(m)}(t,P,Q)$ remainder term, 12

$R^{(m)}(P,Q)$ short form for $R^{(m)}(1,P,Q)$, 14

$r_n = n^{1/2}\, R^{(2)}(P_n,P)$, 99

$r_{W,n} = n^{1/2}\, R^{(2)}(P_{W,n},P_n)$, 99

$R_{W,n} = (r_n + r_{W,n}) / \sigma_{W,n}$, 100

$S^2(P)$ asymptotic variance of $T(P_n)$, 19

S^2 short for $S^2(P)$, 93

$S^{(j)}(.,P)$ j-th canonical gradient of $S(P)$, 21, 62

$S_1(i)$ short for $S^{(1)}(X_i,P)$, 99

$S_n^{\ 2} = E_W\big(n^{1/2}(T(P_{W,n})-T(P_n))\big)^2 / \operatorname{Var}(nW_{1,n})$ estimator of the asymptotic variance, 10, 20

$\tilde{S}(Q)^2$ functional corresponding to an approximation of S_n^2 , 62

$\mathcal{S}_n = n^{1/2}\,(S_n^2 - S^2)/S^2$, 109

Skew(Y) skewness of Y , 33

$\tilde{\mathcal{S}}_n = n^{-1/2}\,\Sigma_{i=1}^n S_1(i)/S^2$, 110

σ^2 scale parameter , 49, 78

σ_n^2 estimator for the scale parameter , 78

$\sigma_{W,n}^2 = \operatorname{Var}(nW_{1,n})$, 19

$\sigma_{W,n}^{(j)^2} = \operatorname{Var}(n\, W_{i,n}^{(j)})$, 33

$\sigma_{Y,n}^2 = \operatorname{Var} Y_{1,n}$, 49

$\sigma_{y,n}^{(j)^2} = \operatorname{Var} Y_{1,n}^{(j)}$, 33

$\hat{\sigma}_n^{\ 2} := n^{-1}\, \Sigma_{i=1}^n\, (X_i - \overline{X}_n)^2$, 5, 49

$\sigma_{\tau,n}^{\ \ 2} := n^{-1}\, \Sigma_{i=1}^n\, \tau_i^{\ 2}$, 56

$\sigma_{\tau,n}^{\ \ 2}(Q)$, 140

$\mathfrak{S}_3$ group of all permutations of 3 elements 63

$T(.)$ functional from $\mathcal{P}$ into $\mathbb{R}$, 1, 9

$T^{(j)}(.,P)$ j-th canonical gradient of $T(.)$ (influence function of order j), 12

$T_1(i)$, $T_2(i,j)$, $T_3(i,j,k)$ short for $T^{(1)}(X_i,P)$, $T^{(2)}(X_i,X_j,P)$ and $T^{(3)}(X_i,X_j,X_k,P)$, 98, 99

$\overline{T}_{1,n} = n^{-1}\Sigma_{i=1}^{n} T_1(i)$, 98

τ_i , $\tau_{i,j}$, $\tau_{i,j,k}$, 56

$\tau^{(i)}(.,Q)$, 140

$\overline{\tau_1^P}$, $\overline{\tau_{1,2}^P}$, $\overline{\tau_1\tau_2\tau_{1,2}}$, 57

Ξ , 139

$u_\alpha = \Phi^{-1}(\alpha)$, 52

$\mathcal{U}_{[0,1]}$ uniform distribution on $[0,1]$,

$U_1, U_2, \ldots, U_n$ i.i.d. $\mathcal{U}_{[0,1]}$, 78

$U_{1,n} \le U_{2,n} \le \ldots \le U_{n,n}$ order statistics of $U_1, U_2, \ldots, U_n$, 85

$U_1^*, U_2^*, \ldots, U_{m_n}^*$ bootstrap sample i.i.d. $\mathbb{U}_n(x)$, 85

$U_{1,m_n}^* \le U_{2,m_n}^* \le \ldots \le U_{m_n,m_n}^*$ order statistics of $U_1^*, U_2^*, \ldots, U_{m_n}^*$, 85

$\mathbb{U}_n(x) := n^{-1}\Sigma_{i=1}^{n} \mathbb{I}\{ U_i \le x \}$ uniform empirical c.d.f., 85

$\mathbb{U}_n^{\leftarrow}$ uniform empirical quantile function, 86

$\mathbb{U}_{W,n}(.) = \Sigma_{i=1}^{n} W_{i,n} \, \mathbb{I}\{ U_i \le x\}$, 86

$\mathbb{U}_{W,n}^{\leftarrow}$ generalized empirical quantile function, 87

$\mathcal{W}_n$ distribution of the weights, 9

$W_{1,n}, \ldots, W_{n,n}$ random weights with distribution $\mathcal{W}_n$, 1, 9

$W_i^{(n)} = nW_{i,n}$, 93

$W_n(i) = nW_{i,n} - 1 = W_i^{(n)} - 1$, 99

$\mathbb{W}_n(t) = \Sigma_{i=1}^{[nt]} W_{i,n}$ exchangeable process , 88

$W_{i,n}^{(1)}$ multinomial weights (Efron's bootstrap) , 32

$W_{i,n}^{(j)} = Y_{i,n}^{(j)} / \Sigma_{i=1}^{n} Y_{i,n}^{(j)}$, j=2,3 simulated i.i.d. generated weights , 32

$X_1, \ldots, X_n$ a sequence of i.i.d. random variables (r.v.) with common probability P , 1

$X_1^*, \ldots, X_n^*$ i.i.d. sample with distribution P_n (bootstrap sample), 1

$\overline{X}_n = n^{-1} \Sigma_{i=1}^{n} X_i$ empirical mean, 4

$\overline{X}_{W,n} = \Sigma_{i=1}^{n} W_{i,n} X_i$ generalized bootstrapped mean, 90

$(Y_{i,n})_{1\le i\le n}$ array of rows i.i.d. random variables, 46

$\overline{Y}_n = n^{-1} \Sigma_{i=1}^{n} Y_{i,n}$, 46

$Y_n(i) = Y_{n,i} - 1$, 129

$Y_{i,n}^{(j)}$ simulated weights , j = 2 mixture of Bernoulli, j = 3 polynomial of N(0,1) , 32

$\mathcal{Y}_n$ distribution of $Y_{i,n}$, 46

AUTHOR INDEX

Abramovitch, L. : 8, 11, 53, 117, 118

Aerts, M. : 7

Akahira, M. : 23

Aldous, D. : 153

Arenal, E. : 177

Arcones, M.A. : 5, 7, 15, 30

Athreya, K.B. : 4, 7, 177

Babu, G.J. : 5, 6

Bahadur, R.R. : 72

Barbe, Ph. : 80, 82, 84, 85, 148, 177

Barndorff-Nielsen, O.E. : 3

Beran, R. : 6, 7, 8, 11, 15, 23, 27, 102

Bertail, P. : 7, 54, 59, 129, 139

Bhattacharya, R.N. : 3, 5, 8

Bickel, P.J. : 4, 5, 6, 7, 8, 10, 15, 107, 108, 114, 129, 139, 186, 188

Billiard, L. : 8

Bingham, N.H. : 175, 178

Birman, M. Š. : 17

Boos, D. : 4

Bretagnolle, J. : 7, 91

Broniatowski, M. : 175

Burke, M.D. : 78

Callaert, H. : 107, 108, 134, 185, 186, 188, 196

Chaganty, N.R. : 73

Chernoff, H. : 72, 74

Chibishov, D. M. : 108, 112

Chow, Y. : 18, 108

Cox, D.R. : 3

Cuesta-Albertos, J.A. : 177

Csörgő, S. : 6, 10, 77, 78, 79, 80, 85, 145, 148, 152, 177

Csörgő, M. : 6, 77, 78, 79, 80, 145, 152

De Haan, L. : 89

Deheuvels, P. : 5, 86, 149, 151, 152

Dehling, H. : 7

Dembo, A. : 142

Denker, M. : 3, 7

Devroye, L. : 32, 34, 51

Diaconis, P. : 4, 153

DiCiccio, T.J. : 4, 8, 27

Dohmann, B. : 6

Dudley, R.M. : 6, 10, 14, 15, 16, 17, 31

Durbin, J. : 78

Dvoretzky, A. : 14

Efron, B. : 2, 3, 4, 7, 8, 9, 10, 26, 27, 31Einmahl, U. : 6, 10, 79, 80, 88, 153, 154

Falk, M. : 4, 6

Feller, W. : 175, 187, 194

Fernholz, L.T. : 12, 14

Field, C. : 3

Filippova, A.A. : 183, 185

Flett, T.M. : 12

Freedman, D.A. : 4, 5, 6, 7, 10, 15, 153

Gaenssler, P. : 6

Gallambos, J. : 89, 149

Ghosh, J.K. : 3

Ghosh, M. : 6, 7, 8, 177

Gill, R.D. : 7, 10, 12, 30

Giné, E. : 5, 6, 7, 10, 15, 30

Gnedenko, B.V. : 194

Goldie, C.M. : 175, 178

Gombay, E. : 78

Gong, G. : 3, 8

Götze, F : 23, 107, 108, 114, 129, 139, 185, 186, 188

Groeneboom, P. : 72, 73

Haeusler, E. : 1, 2, 6, 10, 22, 45, 49, 50, 52, 53, 59, 60, 61, 62, 63

Hahn, M.G. : 10, 77, 154

Hall, P. : 4, 5, 6, 7, 8, 11, 22, 23, 27, 53, 59, 60, 61, 63, 71

Hampel, F. : 13, 15

Härdle, W. : 4

Helmers, R. : 7, 59
Hinkley, D.V. : 8
Hoeffding, W. : 31, 32, 96, 120
Hoffmann-Jørgensen J. : 14
Horváth, L. : 6, 77, 79, 80, 145, 152
Huber, P.J. : 9, 13
Huškova, M. : 7
James, L.F. : 4
Janssen, P. : 7, 107, 108, 134, 185, 186, 188, 196
Karandikar, R.L. : 73
Kaufmann, E. : 6
Kiefer, J. : 14, 149
Knight, K. : 4, 5
Kolmogorov, A.N. : 17, 194
Konishi, S. : 27
Künsch, H.R. : 5
LeCam, L. : 6
LePage, R. : 8
Lincheng, Z. : 130
Liu, R. : 6
Lo, A. Y. : 4, 46
Lohse, K. : 7, 10, 12
Lorentz, G.G. : 17
Low, L.Y. : 7, 177
Mammen, E. : 4, 7, 22
Manski, C.F. : 9
Marron, S. : 4
Martin, M. : 4, 6, 8
Mason, D.M. : 1, 5, 6, 9, 10, 22, 45, 49, 50, 53, 77, 79, 80, 85, 86, 88, 99, 145, 146, 149, 151, 152, 153, 154.
Matran, C. : 177
Mikosch, T. : 7
Millar, P.W. : 6, 7, 15
Monahan, J.F. : 4
Newton, M.A. : 1, 5, 6, 9, 10, 22, 45, 49, 50, 53, 99
Oosterhoff, J. : 72, 73
Parr, W.C. : 6, 7, 8, 10
Pfanzagl, J. : 3, 13, 20, 102, 107

Politis, D.N. : 7, 86, 91
Pollard, D. : 6, 15, 17, 95, 120, 125, 176
Pons, O. : 7, 10, 12
Praestgaard, J. : 6, 9, 18, 22
Quenouille, M.H. : 3
Qumsiyeh, M. : 5, 8
Rachev, S.T. : 14, 15, 31
Rao, R. : 3
Reeds, J.A. : 12, 14, 15, 185
Reid, N. : 3
Reiss R.D. : 4, 6
Révész, P. : 78
Rockafellar, R. : 75
Romano, J.P. : 4, 7, 8, 27, 86, 91
Ronchetti, E. : 3
Rubin, D. : 4, 10
Rudin, W. : 12
Ruymgaart, F.H. : 72, 73
Sen , P.K. : 7, 177
Serfling, J. : 14, 31, 80, 102, 111, 113, 115, 162, 196
Shao, J. : 8
Sheehy, A. : 6, 15
Shorack, G.R. : 6, 86, 149, 151, 152
Silverman B. : 4
Singh, K. : 4, 5, 6, 8, 11, 53, 117, 118
Skorohod, A.V. : 177
Solomjak, M.Z. : 17
Stute, W. : 6
Swanepoel, J.W.H. : 96
Takeuchi, K. : 23
Teicher, H. : 18, 108
Teugels, J. : 175, 178
Tibshirani, R. : 3, 7, 8
Tihomirov, V.M. : 17
Tu, D.S. : 4, 11, 27
Turkheim, E. : 7, 10, 12
Van Zwet, W.R. : 6, 79, 107, 108, 114, 129, 139, 145, 186, 188
Veraverbeke, N. : 7, 107, 108, 134, 185, 186, 188

Von Mises, R. : 7, 9
Weiner, D.C. : 10, 77, 154
Wellner, J.A. : 6, 9, 15, 18, 19, 30, 149
Weng, C.S. : 4
Withers, C.S. : 55, 107
Wolfowitz, J. : 14
Woyczynski, W.A. : 7
Wu, C.F.J. : 8, 9
Yamamuro, H. : 12
Young, G. A. : 4
Zeitouni, O. : 143
Zheng, Z. : 4
Zhidong, B. : 130
Zinn, J. : 5, 6, 10, 15
Zolotarev, V.M. : 14, 45

SUBJECT INDEX

Almost sure consistency : 46, 177

BCAI interval : 27
Bilateral confidence intervals : 29, 51
Bootstrap of the mean : 4, 29, 35, 48, 70, 175, 177
Bootstrapped confidence intervals : 8, 25, 48

Canonical gradients (see influence functions)
Class of functions : 14, 17
Coverage probability : 52, 59, 66
Cramér condition : 23, 138, 184, 185

Differentiability, Fréchet- : 12, 16, 26
Gâteaux- : 12
Hadamard- : 12, 30

Distance : 13, 14

Edgeworth expansion : 23, 53, 135, 138, 183, 197
Edgeworth expansion Inversion : 25
Efficiency : 22
Empirical process : 77, 145, 154
Esseen's lemma : 187
Exchangeable variables : 9, 153
Extreme values : 85

Goodness of fit test : 78

Influence functions : 13, 21, 62, 181

Jackknife : 3, 27

Kullback-Leibler information : 73

Large deviation : 72, 143
Local alternatives : 77

M-estimators : 30

Probability of being censored : 30, 39, 71

Rate of convergence : 17, 95, 103, 119

Second order correctness : 25, 58
Simulations : 32, 70
Small sample asymptotics : 175

Third order correctness : 29, 53

Undersampling : 7, 86, 90
U-statistics : 107, 159

Variance estimation : 8, 20, 163
V-statistics : 31

Lecture Notes in Statistics

For information about Volumes 1 to 12
please contact Springer-Verlag

Vol. 13: J. Pfanzagl (with the assistance of W. Wefelmeyer), Contributions to a General Asymptotic Statistical Theory. vii, 315 pages, 1982.

Vol. 14: GLIM 82: Proceedings of the International Conference on Generalised Linear Models. Edited by R. Gilchrist. v, 188 pages, 1982.

Vol. 15: K.R.W. Brewer and M. Hanif, Sampling with Unequal Probabilities. ix, 164 pages, 1983.

Vol. 16: Specifying Statistical Models: From Parametric to Non-Parametric, Using Bayesian or Non-Bayesian Approaches. Edited by J.P. Florens, M. Mouchart, J.P. Raoult, L. Simar, and A.F.M. Smith, xi, 204 pages, 1983.

Vol. 17: I.V. Basawa and D.J. Scott, Asymptotic Optimal Inference for Non-Ergodic Models. ix, 170 pages, 1983.

Vol. 18: W. Britton, Conjugate Duality and the Exponential Fourier Spectrum. v, 226 pages, 1983.

Vol. 19: L. Fernholz, von Mises Calculus For Statistical Functionals. viii, 124 pages, 1983.

Vol. 20: Mathematical Learning Models — Theory and Algorithms: Proceedings of a Conference. Edited by U. Herkenrath, D. Kalin, W. Vogel. xiv, 226 pages, 1983.

Vol. 21: H. Tong, Threshold Models in Non-linear Time Series Analysis. x, 323 pages, 1983.

Vol. 22: S. Johansen, Functional Relations, Random Coefficients and Nonlinear Regression with Application to Kinetic Data, viii, 126 pages, 1984.

Vol. 23: D.G. Saphire, Estimation of Victimization Prevalence Using Data from the National Crime Survey. v, 165 pages, 1984.

Vol. 24: T.S. Rao, M.M. Gabr, An Introduction to Bispectral Analysis and Bilinear Time Series Models. viii, 280 pages, 1984.

Vol. 25: Time Series Analysis of Irregularly Observed Data. Proceedings, 1983. Edited by E. Parzen. vii, 363 pages, 1984.

Vol. 26: Robust and Nonlinear Time Series Analysis. Proceedings, 1983. Edited by J. Franke, W. Härdle and D. Martin. ix, 286 pages, 1984.

Vol. 27: A. Janssen, H. Milbrodt, H. Strasser, Infinitely Divisible Statistical Experiments. vi, 163 pages, 1985.

Vol. 28: S. Amari, Differential-Geometrical Methods in Statistics. v, 290 pages, 1985.

Vol. 29: Statistics in Ornithology. Edited by B.J.T. Morgan and P.M. North. xxv, 418 pages, 1985.

Vol 30: J. Grandell, Stochastic Models of Air Pollutant Concentration. v, 110 pages, 1985.

Vol. 31: J. Pfanzagl, Asymptotic Expansions for General Statistical Models. vii, 505 pages, 1985.

Vol. 32: Generalized Linear Models. Proceedings, 1985. Edited by R. Gilchrist, B. Francis and J. Whittaker. vi, 178 pages, 1985.

Vol. 33: M. Csörgo, S. Csörgo, L. Horváth, An Asymptotic Theory for Empirical Reliability and Concentration Processes. v, 171 pages, 1986.

Vol. 34: D.E. Critchlow, Metric Methods for Analyzing Partially Ranked Data. x, 216 pages, 1985.

Vol. 35: Linear Statistical Inference. Proceedings, 1984. Edited by T. Calinski and W. Klonecki. vi, 318 pages, 1985.

Vol. 36: B. Matérn, Spatial Variation. Second Edition. 151 pages, 1986.

Vol. 37: Advances in Order Restricted Statistical Inference. Proceedings, 1985. Edited by R. Dykstra, T. Robertson and F.T. Wright. viii, 295 pages, 1986.

Vol. 38: Survey Research Designs: Towards a Better Understanding of Their Costs and Benefits. Edited by R.W. Pearson and R.F. Boruch. v, 129 pages, 1986.

Vol. 39: J.D. Malley, Optimal Unbiased Estimation of Variance Components. ix, 146 pages, 1986.

Vol. 40: H.R. Lerche, Boundary Crossing of Brownian Motion. v, 142 pages, 1986.

Vol. 41: F. Baccelli, P. Brémaud, Palm Probabilities and Stationary Queues. vii, 106 pages, 1987.

Vol. 42: S. Kullback, J.C. Keegel, J.H. Kullback, Topics in Statistical Information Theory. ix, 158 pages, 1987.

Vol. 43: B.C. Arnold, Majorization and the Lorenz Order: A Brief Introduction. vi, 122 pages, 1987.

Vol. 44: D.L. McLeish, Christopher G. Small, The Theory and Applications of Statistical Inference Functions. vi, 124 pages, 1987.

Vol. 45: J.K. Ghosh (Editor), Statistical Information and Likelihood. 384 pages, 1988.

Vol. 46: H.-G. Müller, Nonparametric Regression Analysis of Longitudinal Data. vi, 199 pages, 1988.

Vol. 47: A.J. Getson, F.C. Hsuan, {2}-Inverses and Their Statistical Application. viii, 110 pages, 1988.

Vol. 48: G.L. Bretthorst, Bayesian Spectrum Analysis and Parameter Estimation. xii, 209 pages, 1988.

Vol. 49: S.L. Lauritzen, Extremal Families and Systems of Sufficient Statistics. xv, 268 pages, 1988.

Vol. 50: O.E. Barndorff-Nielsen, Parametric Statistical Models and Likelihood. vii, 276 pages, 1988.

Vol. 51: J. Hüsler, R.-D. Reiss (Editors), Extreme Value Theory. Proceedings, 1987. x, 279 pages, 1989.

Vol. 52: P.K. Goel, T. Ramalingam, The Matching Methodology: Some Statistical Properties. viii, 152 pages, 1989.

Vol. 53: B.C. Arnold, N. Balakrishnan, Relations, Bounds and Approximations for Order Statistics. ix, 173 pages, 1989.

Vol. 54: K.R. Shah, B.K. Sinha, Theory of Optimal Designs. viii, 171 pages, 1989.

Vol. 55: L. McDonald, B. Manly, J. Lockwood, J. Logan (Editors), Estimation and Analysis of Insect Populations. Proceedings, 1988. xiv, 492 pages, 1989.

Vol. 56: J.K. Lindsey, The Analysis of Categorical Data Using GLIM. v, 168 pages, 1989.

Vol. 57: A. Decarli, B.J. Francis, R. Gilchrist, G.U.H. Seeber (Editors), Statistical Modelling. Proceedings, 1989. ix, 343 pages, 1989.

Vol. 58: O.E. Barndorff-Nielsen, P. Blæsild, P.S. Eriksen, Decomposition and Invariance of Measures, and Statistical Transformation Models. v, 147 pages, 1989.

Vol. 59: S. Gupta, R. Mukerjee, A Calculus for Factorial Arrangements. vi, 126 pages, 1989.

Vol. 60: L. Györfi, W. Härdle, P. Sarda, Ph. Vieu, Nonparametric Curve Estimation from Time Series. viii, 153 pages, 1989.

Vol. 61: J. Breckling, The Analysis of Directional Time Series: Applications to Wind Speed and Direction. viii, 238 pages, 1989.

Vol. 62: J.C. Akkerboom, Testing Problems with Linear or Angular Inequality Constraints. xii, 291 pages, 1990.

Vol. 63: J. Pfanzagl, Estimation in Semiparametric Models: Some Recent Developments. iv, 112 pages, 1990.

Vol. 64: S. Gabler, Minimax Solutions in Sampling from Finite Populations. v, 132 pages, 1990.

Vol. 65: A. Janssen, D.M. Mason, Non-Standard Rank Tests. vi, 252 pages, 1990.

Vol. 66: T. Wright, Exact Confidence Bounds when Sampling from Small Finite Universes. xvi, 431 pages, 1991.

Vol. 67: M.A. Tanner, Tools for Statistical Inference: Observed Data and Data Augmentation Methods. vi, 110 pages, 1991.

Vol. 68: M. Taniguchi, Higher Order Asymptotic Theory for Time Series Analysis. viii, 160 pages, 1991.

Vol. 69: N.J.D. Nagelkerke, Maximum Likelihood Estimation of Functional Relationships. v, 110 pages, 1992.

Vol. 70: K. Iida, Studies on the Optimal Search Plan. viii, 130 pages, 1992.

Vol. 71: E.M.R.A. Engel, A Road to Randomness in Physical Systems. ix, 155 pages, 1992.

Vol. 72: J.K. Lindsey, The Analysis of Stochastic Processes using GLIM. vi, 294 pages, 1992.

Vol. 73: B.C. Arnold, E. Castillo, J.-M. Sarabia, Conditionally Specified Distributions. xiii, 151 pages, 1992.

Vol. 74: P. Barone, A. Frigessi, M. Piccioni (Editors), Stochastic Models, Statistical Methods, and Algorithms in Image Analysis. vi, 258 pages, 1992.

Vol. 75: P.K. Goel, N.S. Iyengar (Editors), Bayesian Analysis in Statistics and Econometrics. xi, 410 pages, 1992.

Vol. 76: L. Bondesson, Generalized Gamma Convolutions and Related Classes of Distributions and Densities. viii, 173 pages, 1992.

Vol. 77: E. Mammen, When Does Bootstrap Work? Asymptotic Results and Simulations.vi, 196 pages, 1992.

Vol. 78: L. Fahrmeir, B. Francis, R. Gilchrist, G. Tutz (Editors), Advances in GLIM and Statistical Modelling: Proceedings of the GLIM92 Conference and the 7th International Workshop on Statistical Modelling, Munich, 13-17 July 1992. ix, 225 pages, 1992.

Vol. 79: N. Schmitz, Optimal Sequentially Planned Decision Procedures. xii, 209 pages, 1992.

Vol. 80: M. Fligner, J. Verducci (Editors), Probability Models and Statistical Analyses for Ranking Data. xxii, 306 pages, 1992.

Vol. 81: P. Spirtes, C. Glymour, R. Scheines, Causation, Prediction, and Search. xxiii, 526 pages, 1993.

Vol. 82: A. Korostelev and A. Tsybakov, Minimax Theory of Image Reconstruction. xii, 268 pages, 1993.

Vol. 83: C. Gatsonis, J. Hodges, R. Kass, N. Singpurwalla (Editors), Case Studies in Bayesian Statistics. xii, 437 pages, 1993.

Vol. 84: S. Yamada, Pivotal Measures in Statistical Experiments and Sufficiency. vii, 129 pages, 1994.

Vol. 85: P. Doukhan, Mixing: Properties and Examples. xi, 142 pages, 1994.

Vol. 86: W. Vach, Logistic Regression with Missing Values in the Covariates. xi, 139 pages, 1994.

Vol. 87: J. Møller, Lectures on Random Voronoi Tessellations.vii, 134 pages, 1994.

Vol. 88: J.E. Kolassa, Series Approximation Methods in Statistics.viii, 150 pages, 1994.

Vol. 89: P. Cheeseman and R.W. Oldford (Editors), Selecting Models From Data: Artificial Intelligence and Statistics IV. x, 487 pages, 1994.

Vol. 90: A. Csenki, Dependability for Systems with a Partitioned State Space: Markov and Semi-Markov Theory and Computational Implementation. x, 241 pages, 1994.

Vol. 91: J.D. Malley, Statistical Applications of Jordan Algebras. viii, 101 pages, 1994.

Vol. 92: M. Eerola, Probabilistic Causality in Longitudinal Studies. vii, 133 pages, 1994.

Vol. 93: Bernard Van Cutsem (Editor), Classification and Dissimilarity Analysis. xiv, 238 pages, 1994.

Vol. 94: Jane F. Gentleman and G.A. Whitmore (Editors), Case Studies in Data Analysis. viii, 262 pages, 1994.

Vol. 95: Shelemyahu Zacks, Stochastic Visibility in Random Fields. x, 175 pages, 1994.

Vol. 96: Ibrahim Rahimov, Random Sums and Branching Stochastic Processes. viii, 195 pages, 1995.

Vol. 97: R. Szekli, Stochastic Ordering and Dependence in Applied Probablility, viii, 194 pages, 1995.

Vol. 98: Philippe Barbe and Patrice Bertail, The Weighted Bootstrap, viii, 230 pages, 1995.

General Remarks

Lecture Notes are printed by photo-offset from the master-copy delivered in camera-ready form by the authors of monographs, resp. editors of proceedings volumes. For this purpose Springer-Verlag provides technical instructions for the preparation of manuscripts. Volume editors are requested to distribute these to all contributing authors of proceedings volumes. Some homogeneity in the presentation of the contributions in a multi-author volume is desirable.

Careful preparation of manuscripts will help keep production time short and ensure a satisfactory appearance of the finished book. The actual production of a Lecture Notes volume normally takes approximately 8 weeks.

For monograph manuscripts typed or typeset according to our instructions, Springer-Verlag can, if necessary, contribute towards the preparation costs at a fixed rate.

Authors of monographs receive 50 free copies of their book. Editors of proceedings volumes similarly receive 50 copies of the book and are responsible for redistributing these to authors etc. at their discretion. No reprints of individual contributions can be supplied. No royalty is paid on Lecture Notes volumes.

Volume authors and editors are entitled to purchase further copies of their book for their personal use at a discount of 33.3% and other Springer mathematics books at a discount of 20% directly from Springer-Verlag. Authors contributing to proceedings volumes may purchase the volume in which their article appears at a discount of 20%.

Springer-Verlag secures the copyright for each volume.

GPSR Compliance

The European Union's (EU) General Product Safety Regulation (GPSR) is a set of rules that requires consumer products to be safe and our obligations to ensure this.

If you have any concerns about our products, you can contact us on ProductSafety@springernature.com

In case Publisher is established outside the EU, the EU authorized representative is:

Springer Nature Customer Service Center GmbH
Europaplatz 3
69115 Heidelberg, Germany

Batch number: 10583998

Printed by Printforce, the Netherlands